London School of Economics Industrial Relations Series
General Editor: Professor B. C. Roberts

RELUCTANT MILITANTS

Reluctant Militants:

A Study of Industrial Technicians

B. C. ROBERTS
RAY LOVERIDGE
JOHN GENNARD
J. V. EASON
and others

HEINEMANN EDUCATIONAL BOOKS
LONDON

Heinemann Educational Books Ltd

LONDON   EDINBURGH   MELBOURNE   TORONTO
AUCKLAND   JOHANNESBURG   SINGAPORE
IBADAN   NAIROBI   HONG KONG   NEW DELHI
KUALA LUMPUR

ISBN 0 435 85780 0

© B. C. Roberts, Ray Loveridge and John Gennard 1972

First published 1972

Published by Heinemann Educational Books Ltd
48 Charles Street, London W1X 8AH

Printed in Great Britain by Willmer Brothers Limited, Birkenhead

Contents

Contents

Preface

The researching and the writing of this book was a co-operative effort by members of the Department of Industrial Relations at the London School of Economics. The three authors were greatly assisted by Joe Eason, whose contribution to the field work was very great indeed. Others who gave their assistance at different stages were Nicholas Bosanquet, Keith Murphy, Robert Banks, Gill Palmer, Margaret Attwood and Sheila Rothwell.

In carrying out research in the field of industrial relations the willingness of trade unions and firms to co-operate is of vital importance. It is not possible to say that the trade unions concerned gave enthusiastic support to this investigation, but they at least did not make it impossible. We are extremely grateful to individual trade union officers and members who were extremely helpful and contributed essential information. A good number of the companies approached for permission to interview their technicians refused, fearing that the presence of an interviewer might have an unfortunate effect on their industrial relations. However, fifteen companies were not deterred by these considerations and gave us a great deal of help without which it would have been impossible to make this study. To their managers and workers we offer our deepest thanks.

Empirical studies inevitably cost money. The financial assistance of the Social Science Research Council was an essential element in the making of this study. We are deeply grateful for the help that we received in this respect.

B. C. ROBERTS
RAY LOVERIDGE
JOHN GENNARD
February 1972

PART I

PART 1

I Changing Patterns of Demand and Supply: Some Causes and Some Effects

During this century technological change has led to a growth in the demand for scientific and technical manpower. Technological change is difficult to define and its exact course of development in the future can only be guessed, but it would seem safe to predict that the growth in demand for scientific and technical manpower is likely to continue. Before more precise predictions of manpower needs and the kind of economic and social problems that are likely to emerge in the future can be made it will be necessary to develop effective ways of measuring technological change, and the impact of technology on diverse and disparate social cultures.

So far all the economists' measures of technological change have suffered from serious drawbacks in that, while having some analytical merit, they are in practice extremely difficult to apply.[1] Nevertheless, although it cannot be easily and accurately measured, the pace of technological change is continuing to increase, and it seems certain to continue to do so in the future. Technological change affects both the level and the structure of employment, and thus affects the patterns of social behaviour. In this study attention is focused on one group of employees that are intimately linked with the course of technological change, namely, those known as *technicians*. The definition of a technician is by no means a simple matter, as will be shown later.

[1] See for example, E. Domar, 'On the Measurement of Technological Change', *Economic Journal*, December 1961. A review of the whole problem is provided by M. Brown, *On the Theory and Measurement of Technological Change* (Cambridge University Press, 1966).

The functions which the technician carries out, and the role that he plays in the labour force are, while changing, likely to grow rather than to diminish in importance.

Technological change and the pattern of demand for labour

Technological change affects both the level and the structure of employment. This structure is normally seen in occupational terms. At the top are usually placed administrative, technical and clerical jobs. These occupations have historically demanded a relatively high level of formal education and have included the highest-paid jobs in a company, industry or economy; for example, managers and technologists. Skilled and semi-skilled manual occupations are usually placed at the lower end of the structure, and the unskilled workers at the very bottom. In the centre, between, but overlapping the technologists and skilled manual workers, is the occupational group which is the concern of this study, the technicians.

Technological change affects the type of skills required. Some existing jobs are 'destroyed' or modified and new types of jobs may be created. The modification of existing jobs may influence the degree to which the currently trained labour force moves between occupations. The introduction of new techniques and changes in skill requirements affect the employer's ability to substitute one type of technician for another or for a technologist; or, to substitute a manual worker for a technician. The implications of changes in the elasticity of substitution between technicians and technologists and technicians and manual workers will be elaborated throughout this study.

A major difficulty in analysing the impact of technological change on occupational structure is that of isolating the pure effect of technological change. A whole range of influences are at work moulding the occupational structure at any point in time; for example, changes in consumer demand, and growth in the size of the units of employment. Many of these forces are interdependent. An increase in the size of the business organizations *per se* may affect the degree of specialization and other aspects of the occupational structure within a plant, yet it may be that the existing level of technology sets constraints on the minimum viable unit of business organization and therefore effectively determines the area of *choice* open to the entrepreneur. Even these constraints may exist only so long as human imagination recognizes them as such. For this reason a major element in modern capitalism, whether private or state-owned, has been the competitive importance attached to innovation.[1]

This concern with emulative innovation is often manifested at governmental level both in Britain and elsewhere. Harold Wilson's 1964 election promise of a 'white-hot technical revolution' resulted in the creation of a Ministry of Technology designed to stimulate Research and Development and a greater emphasis on technological investment within industry. Of these two activities the latter has come to be seen as being the more important. The paucity of Government effort given to increasing the demand for new research had been the subject of much criticism.[2] However, the publication of the Brookings Report in 1968,[3] and in particular Professor M. J. Peck's contribution,[4] appeared to go some way towards justifying this lack of initiative.

Professor Peck has suggested that 'Britain's scientific and technical resources have been and are now overcommitted'. The resource with which he is mainly concerned is that of manpower, or more particularly, scientists and engineers. The table reproduced below provides the starting point for his argument. This sets out the ratio of Research and Development spending to gross national product in U.S. prices and the proportion of the labour force that possess high technical qualifications.

Britain is spending over half as much again on Research and Development as the U.S.A. relative to her G.N.P., but with a much smaller proportion of the working population employed as scientists and technologists. Of itself this table does not provide evidence to suggest that the occupational mix used by British industry may be a less efficient one than that used in other countries (in the sense that the U.K. uses a cheaper labour input). Professor Peck is, however, able to point to the high concentration of demand for qualified scientists in particular industrial labour markets. British Defence expenditure, for example,

---

[1] This thesis has been most fully developed by Joseph A. Schumpeter, *Capitalism, Socialism and Democracy* (Allen and Unwin, London 1961).

[2] See for example 'Technology: The Revolution That Wasn't', *The Economist*, 19 March 1966, p. 1147.

[3] Richard E. Caves *et al.*, *Britain's Economic Prospects*, a Brookings Institution Study (Allen and Unwin, London 1968).

[4] Op. cit. Part X, Merton J. Peck, 'Science and Technology', pp. 448–484.

Table 1.1   Research and development in relation to available
scientific manpower, 1962

Country	Ratio of total R. & D. expenditures to G.N.P. (U.S. = 100)	Scientists and engineers as percentage of employment, 1959
United States	100	1.7
United Kingdom	151	1.0
France	124	0.8
Germany	65	1.3
Netherlands	115	1.0
Belgium	70	1.0

SOURCE: R. E. Caves (ed.) Table 10-1, page 449.
Britain's Economic Prospects.

is the third largest in the world, and so is that of the aircraft
industry. In fact the demand for qualified technical staff is shown
to be highly concentrated within a small group of industries which
account for an increasing share of the demand for manufactured
goods, and in particular an increasing proportion of international
trade. This group consists of aircraft, electronics, other electrical,
chemicals and allied products, machinery, vehicles and instruments. Professor Peck sees Britain's over-commitment as being in
the area of pure research rather than in technology or applied
research.[1] The ratio of engineers to scientists employed in British
industry (54:46) is very much lower than in the U.S.A. (71:29)
or other European nations. This disproportionate allocation of
labour takes place within a much lower scientific population,
leaving Britain with the lowest proportion of technically qualified
engineers relative to the rest of the work force. The corollary to
this is of course that Britain employs relatively more pure
scientists.

In part this might be accounted for by the nature of the
research function in British industry (i.e. the emphasis on pure
research). However, evidence suggests that the most pervasive
solution to the lack of engineering expertise discovered by British

[1] This suggestion is not entirely new. The *Report of the Committee on Scientific
Manpower* (Barlow) in 1946 argued that 'whereas no one can doubt the value of
our achievements in fundamental science during the war, we are not always
successful in those applications of science which lie in the field of engineering
and technology' (p. 10).

Table 1.2 Percentage distribution of engineers and scientists and proportion in industry, five countries, 1959

Country	Engineers	Natural scientists	Agricultural scientists	Percentage of total scientists and engineers in industry
United Kingdom	54	43	3	42
United States	71	25	4	74
Belgium	77	14	9	54
France	67	21	12	52
Netherlands	80	12	7	n.a.

SOURCE: O.E.C.D., *Resources of Scientific and Technical Personnel in the O.E.C.D. Area*, pp. 113, 117.

employers is the substitution of the 'non-professional for the professional', or the technician for the technologist. Overall British industry employs 4.7 technicians per professional as against the American ratio of 0.62 technicians per professional (see Table 1.3).

In the compilation of all of these international comparisons

Table 1.3 Qualified engineers, scientists and technicians as percentage of total employees by industry, United Kingdom and United States, 1959

Industry	United Kingdom			United States		
	Engineers and Scientists	Technicians	Total	Engineers and Scientists	Technicians	Total
Electrical engineering	2.2	12.3	14.5	7.6	5.5	13.1
Chemical and mineral oil refining	3.9	9.0	12.9	7.9	3.3	11.2
Plant and machinery	1.4	9.1	10.5	4.2	4.1	8.3
Metal manufacture	0.8	4.5	5.3	2.6	1.3	3.9

SOURCE: Advisory Council on Scientific Policy, Committee on Scientific Manpower, *The Long-Term Demand for Scientific Manpower*, Cmnd 1490, H.M.S.O., London 1961).

there are obviously problems of definition.[1] The British 'technician' might be closer in skill to the American 'engineer', for example, so that differences in title disguise similar inputs of skills and training. Yet even if the holders of the British Higher National Certificate, a non-graduate qualification, are regarded as engineers, the British still have two technicians for each of their fewer and scarcer professional technologists compared with 0.62 technicians per professional engineer in the U.S.A. The Feilden Committee also commented strongly on the low standard of design in British engineering after the initial basic research and design process.[2] This and other evidence does not support the hypothesis that these international differences in occupational weighting are mere matters of definitional semantics.

It may also be argued that the substitution of technologists may be justified in terms of the supply price of professional engineers as against less qualified technicians. In Britain technicians receive approximately 60 per cent of the professional engineer's salary; in the United States they receive 68 per cent. In terms of starting salaries and employers' own estimates of demand, engineers in Britain have been improving their relative position over the past decade compared to that of their U.S. counterparts. Yet it appears unlikely that the difference in the salaries of technicians and engineers would bring about the contrasts in the allocation of resources indicated by the differences in manpower ratios; one has perhaps to look outside market factors to account for international differences in the occupational structure.

Despite increases in their salaries, there has been little change in the flow of professional engineers into industry up to very recently. One might, therefore, seek an explanation in the long-term evolution of the structure of technical employment in the U.K. and in the institutions developed to meet the needs for trained manpower. Certainly it appears that, faced with quite similar technological constraints, employers, governments and educationalists in Britain have chosen to structure their organizations and to motivate their labour force in a different way from that adopted in other countries.

---

[1] These are discussed at greater length in Chapter 2.

[2] Report of a Committee appointed by the Council for Scientific and Industrial Research (Chairman, G. B. R. Feilden), *Engineering Design* (H.M.S.O., London 1963).

The long-term pattern of demand

An occupational hierarchy represents a normative designation of each occupational situation in an order accorded to it with respect to a number of characteristics. Important among these are the operative skills required to perform the occupant's job and/or the authority allocated to his role within the place of work. Earnings provide another index, but perhaps more important in the compilation of many indices is the differential allocation of prestige within wider society. The problem of determining the influence of each of these factors upon the definition of 'an occupation' is one of the major themes of this book. But each of the indices represents problems in definition.

Skill, for example, involves different degrees of knowledge, responsibility, physical capability and experience; the relative amounts of each factor contained within a job specification will be influenced over time by a changing socio-technical environment. But these elements are not always capable of quantitative measurement in that they are sometimes unique and therefore cannot be aggregated. Because of this it is necessary to treat with caution such beliefs as that the level of skill in the labour force must be increasing since an increasing proportion of people are now entering non-operative jobs. One way out of the difficulty has been to refer to skill only in connection with manual workers, and measure it by the length and type of training, for example apprenticeships. This is open to complications with the growth of alternative training schemes of a qualitatively different nature, but perhaps more important, it confuses the employer's demand function with his supply function, i.e. what is available to him in the market by way of 'trained' labour. Some writers[1] have equated skill with high wages, but this substitution of one interdependent variable for another is tautological and does not carry one any further towards objective measurement. Even the mixture of factors making up the concept of skill is difficult to define accurately.[2]

Bearing these limitations of measures in mind, it is possible to examine existing statistics recounting the changes in the pat-

[1] K. G. J. C. Knowles and D. J. Robertson, 'Differences in Wages between Skilled and Unskilled Workers 1880–1950', *Bulletin of the Oxford University Institute of Economics and Statistics* Vol. 13 No. 2, (April 1951).
[2] See for example R. L. Raimon, 'Changes in Productivity and Skill Mix', *International Labour Review*, Vol. 92, No. 4, (October 1965).

tern of demand for manpower in the U.K. over time. The paucity
of data makes it necessary to confine the examination to the
period 1911 to the 1960s. Occupational statistics for the U.K.
have only appeared with regular frequency since the Ministry
of Labour first undertook a survey of occupations in manufacturing in May 1963, and published the results in its December 1963
Gazette: this survey is now an annual feature.[1] It crudely
divides occupations into four main broad groups—Administrative, Technical and Clerical; Skilled Manual Occupations; Semi-
skilled Occupations; and other employees. Prior to this the only
regular collections of statistics on occupations were those published in the Population Census. The Registrar-General uses a
social and socio-economic classification of occupations that has a
different sample frame from that used in the Ministry of Labour
surveys. The full census is only taken every ten years, leaving long
time-gaps, but matters have been made worse by the fact that
each census has used a different occupational classification.

Occupational classifications have to be modified from time to
time to take account of changes in job content and other factors.
This makes comparisons over industries, countries or long periods
of time extremely difficult to make. In an endeavour to overcome
these problems, long-term studies of change[2] have to devise
modifications of the system of classification used in official sources.
Yet the accepted ordering of the hierarchy is slow to change.
It remains, for example, usual to consider the developing
structure of the employed labour force in terms of two main
categories—white-collar and manual workers. Traditionally the
term 'white-collar worker' has been applied to most workers in
jobs which did not involve heavy physical exertion and which
were performed in surroundings that were clean, stable and
relatively free from hazards. This dichotomy reflects a broader
social evaluation which does not always coincide with the
actual job conditions, either because the latter have changed over
time or because the 'image' of the job possessed by the general

[1] However, the 1969 survey was confined to the Metal Trades Industries
Order Classifications (V–IX), and this is to be the case for all future surveys.
See *Employment and Productivity Gazette*, January 1970.

[2] For example, G. Routh, *Occupation and Pay in Great Britain 1906–60*
(Cambridge University Press, 1965) and G. S. Bain, 'The Growth of White-
Collar Unionism in Great Britain', *British Journal of Industrial Relations*,
Vol. IV, No. 3 (November 1966). See also *The Growth of White Collar Unionism*
(Oxford University Press 1970). The use of regression analysis in the latter
work must be treated with some caution for all of the reasons given above.

public reflects the occupants' proximity to that of a high status group.[1] Technicians may, for example, be doing extremely routine, dirty and even dangerous work, yet in their own eyes and those of others it remains superior to that of a so-called 'manual worker'.

The growth in the number of white-collar workers in the U.K. has been both absolute and relative (see Table 1.4) and the general consensus appears to be that the trend will not only continue but in some instances be exponential. It has been predicted that by 1980 the U.K. will have more of its working population in white-collar jobs than in manual occupations.[2] Between 1911 and 1961 white-collar occupations increased by 147 per cent while manual occupations grew by only 2 per cent. During the same period the total occupied population increased by 29 per cent. The number of manual workers has declined in absolute numbers since 1931 from 14,776,000 to 14,020,000. In 1911 white-collar workers accounted for 18.7 per cent of the total occupied population, but by 1961 this figure had risen to 35.9 per cent. Manual workers as a proportion of the active working population have declined from 74.6 per cent in 1911 to 59.3 per cent in 1961.

The growth of demand for scientific and technical manpower

Table 1.4 shows that the growth rate in numbers of higher professionals relative to that of lower professionals and technicians supports Peck's hypothesis. A closer examination of published census data shows that although the number employed in scientific and technical occupations is small it does represent the fastest-growing element within the white-collar occupations. It was not until the 1921 Census of Population that draughtsmen and laboratory assistants were considered sufficiently important groups to merit a separate classification. Between 1921 and 1961 the number of scientists and engineers grew by 688 per cent, the number of draughtsmen by 376 per cent and the number of laboratory assistants by 1,820 per cent. The Ministry of Labour Manpower Research Unit,[3] in reviewing

[1] See for example the description of the nineteenth-century clerk in D. Lockwood, *The Black-coated Worker* (Allen and Unwin, London 1958).
[2] G. S. Bain, *The Growth of White Collar Unionism* (Oxford University Press, 1970).
[3] Ministry of Labour, *Occupational Changes 1951–61*, Manpower Studies No. 6 (H.M.S.O., London 1967).

Table 1.4 The occupied population of Great Britain by major occupational groups

Occupational groups	Number of persons in major occupational groups (1911–1961) in thousands					Major occupational groups as a percentage of total occupied population (1911–1961)					Growth indices of major occupational groups (1911–1961) 1911 = 100				
	1911	1921	1931	1951	1961	1911	1921	1931	1951	1961	1911	1921	1931	1951	1961
1. Employers and proprietors	1,232	1,318	1,407	1,117	1,139	6.7	6.8	6.7	5.0	4.7	100	107	114	91	92
2. All white-collar workers	3,433	4,094	4,842	6,948	8,480	18.7	21.2	23.0	30.9	35.9	100	119	141	202	247
(a) Managers and administrators	631	704	770	1,245	1,268	3.4	3.6	3.7	5.5	5.4	100	112	122	197	201
(b) Higher professionals	184	196	240	435	718	1.0	1.0	1.1	1.9	3.0	100	107	130	236	390
(c) Lower professionals and technicians	560	679	728	1,059	1,418	3.1	3.5	3.5	4.7	6.0	100	121	130	189	253
(d) Foremen and inspectors	237	279	323	590	682	1.3	1.4	1.5	2.6	2.9	100	118	136	249	288
(e) Clerks	832	1,256	1,404	2,341	2,996	4.5	6.5	6.7	10.4	12.7	100	151	169	281	360
(f) Salesmen and shop assistants	989	980	1,376	1,278	1,398	5.4	5.1	6.5	5.7	5.9	100	99	139	129	141
3. All manual workers	13,685	13,920	14,776	14,450	14,020	74.6	72.0	70.3	64.2	59.3	100	102	108	106	102
4. Total occupied population	18,350	19,332	21,024	22,515	23,639	100	100	100	100	100	100	105	115	123	129

SOURCE: G. S. Bain, 'The Growth of White-Collar Unionism in Great Britain,' *British Journal of Industrial Relations*, Vol. IV, No. 3 (November 1966),

occupational changes between 1951 and 1961, estimated that of the growth in professional and technical occupations, a little under a quarter was accounted for by expansion in scientists, engineers and technologists taken together, and a further quarter by the growth in the employment of industrial technicians. These two groups grew faster than other non-manual occupations in relation to employment. However, the demand for technicians has been increasing at a more rapid rate than that for technologists.

Table 1.5 shows the growth in the number of scientists and technologists, and in the group which is the particular concern of this study, draughtsmen and other technicians employed in manufacturing industry, in 1964 and 1968. The figures are taken from the Ministry of Labour's annual survey of occupations in manufacturing industry. The total number of scientists and technologists increased from 75,710 to 94,850, an increase of 27.9 per cent; technicians increased by 38 per cent. From these figures it appears that the exponential growth rate shown for both 'higher professionals' and 'lower professionals and technicians' in Table 1.4 may be continuing. The average growth rate among scientists and technologists over the four years 1964–68 was 7 per cent, as against the 5 per cent average annual rate shown by the former group over the fifty years covered in Table 1.4. That for technicians was 9.5 per cent, as against the 7.5 per cent shown by the latter group over the same period. This implies that in aggregate terms, the difference in numbers between the two groups was at that time getting greater despite evidence that the rate of growth in numbers of technicians has declined somewhat in recent years.

In some industries the technician labour force was *considerably* larger than the technologist element. The Ministry of Labour's survey of Metal Industries showed that in 1963 the firms covered employed rather more than two technicians to one scientist or technologist. When draughtsmen were included, the technician/technologist ratio rose to 4 : 1.[1] A survey of the employment of technicians in the Chemical and Engineering Industries showed the ratio 5 : 1.[2]

---

[1] Ministry of Labour, *The Metal Industries*, Manpower Studies No. 2 (H.M.S.O., London 1965), p. 15.
[2] Ministry of Labour, 'Survey of the Employment of Technicians in the Chemical and Engineering Industries', *Ministry of Labour Gazette*, December 1960.

Table 1.5 Changes in the number of technical staff between 1964 and 1968, by Standard Industrial Classification

Standard Industrial Classification²	Scientists and technologists			Draughtsmen			Other technicians		
	1964	1968	Percentage change 1968 over 1964	1964	1968	Percentage change 1968 over 1964	1964	1968	Percentage change 1968 over 1964
Food, drink and tobacco	2,770	2,980	+ 7.6	1,150	1,290	+ 12.2	3,640	5,730	+ 57.4
Chemical and allied industries	7,100	17,280	+ 1.1	3,010	2,440	− 18.9	22,920	27,060	18.1
Metal manufacture	4,950	5,650	+ 14.2	4,440	4,210	− 5.2	10,240	14,150	+ 38.2
Engineering and electrical goods	33,200	44,920	+ 35.3	65,160	64,380	− 1.2	58,230	89,200	+ 53.2
Shipbuilding and ship-repairing	—	870	—	—	3,540		—	1,780	—
Marine engineering	220	620	+180.2	1,950	1,360	− 30.3	700	1,040	+ 48.6
Vehicles	7,000	13,000	+ 85.7	19,820	16,030	− 19.1	24,980	32,370	+ 29.6
Manufacture of metal goods	1,610	1,650	+ 2.5	5,430	4,670	− 14.0	3,070	5,030	+ 63.9

Textiles	3,000	2,760	− 8.0	1,580	590	− 62.7	7,100	6,690	− 5.8
Leather, leather goods and fur	150	30	− 80.0	10	—	—	201	190	− 9.5
Clothing	120	170	+ 42.7	220	90	− 59.1	1,410	1,180	− 16.3
Footwear	80	80	+ 12.5	90	70	− 22.2	620	720	+ 16.1
Bricks, glass and cement	1,680	1,770	+ 5.4	2,120	2,000	− 5.7	2,170	3,450	+ 59.0
Pottery	120	150	+ 24.7	200	150	− 25.0	370	390	+ 5.4
Timber and furniture	50	150	+200.0	1,420	1,610	+ 11.8	760	950	+ 25.0
Paper, paperboard and cardboard	670	700	+ 4.5	490	490	—	2,300	2,220	− 3.5
Printing and publishing	890	390	− 56.2	700[1]	700[1]	+164.3	320	1,060	+231.3
Other manufacturing	2,080	1,890	− 9.1	1,800	1,440	− 20.0	3,740	3,930	+ 5.1
Total manufacture	75,710	94,850	+ 27.9	109,590	106,210	− 3.1	142,870	197,130	+ 38.0

NOTES: 1. Includes designers and typographers.   SOURCE: Ministry of Labour, 'Annual Survey of Occupations in Manufacturing',
2. As revised in 1958.   May 1964, May 1968, *Ministry of Labour Gazette*, January 1965,
January 1969.

Yet the growth in the employment of technologists and technicians has not been uniformly spread across manufacturing industries. Their numbers in 'declining' or 'rationalizing' industries shown in Table 1.5 actually dwindled or stood still over the period 1964–68. In others, such as Marine Engineering, Timber and Furniture and Pottery, the numbers of scientists and technologists increased faster than those of technicians. Differences such as these may reflect part of the 'catching up' process in which more traditional industries bring in new techniques and product changes in their endeavour to retain or expand their markets in a world in which the pace of change is being set by more recently established industries. But in general the growth rate of 'Other Technicians' is far greater than that of scientists and technologists, even in those industries such as shipbuilding and ship-repairing, metal manufacture, engineering and electrical goods, and chemical and allied industries, which were first to employ technologists and technicians and which have displayed some slowing down in their capacity to absorb technical staff.

Occupational differences within the umbrella definition of 'technician' were also to be seen and were most clearly distinguishable among draughtsmen. The numbers of draughtsmen fell by 3 per cent between 1964 and 1968, and there were indications that this was part of a longer-term decline within this group. In 1964 there were 109,590 draughtsmen employed in manufacturing but by 1968 the figure had fallen to 106,210. Again there were large variations between industries, but two of Britain's leading growth industries, vehicles and chemicals, both reported reductions in the number of draughtsmen employed of 19 per cent.

One possible explanation for the absolute decline in the number of draughtsmen employed has been the growth of computer design techniques. The Ministry of Labour's Manpower Research Unit Report on the Metal Industries drew attention to the trend in the employment of draughtsmen relative to other technicians. 'Advancing computer techniques and the increased influence of scientists and technologists in design may be among the factors accounting for this.'[1]

In the survey upon which this study is based there was little evidence to suggest that computerized techniques were making draughtsmen redundant. There appeared much more reason to

[1] Ministry of Labour, *The Metal Industries*, Manpower Studies No. 2, (H.M.S.O., London 1965), p. 15.

see 'the increased influence of scientists and technologists' at work in so far as the laboratory and the photographer's dark-room were becoming centres of design, rather than the drawing office. In general it seems impossible to say that one or even a small number of technological changes have, of themselves, brought about the decline of draughting. Rather it seems more likely that the design function is becoming increasingly dispersed over a whole range of more specialized jobs, which range from design engineer to production planning. It is extremely rare for a machine to directly take over a draughtsman's role; the demise of the latter is usually accompanied by the creation of several other roles at various levels of expertise but usually demanding skills that are much more specific to the work situation than the more generalized craft accomplishments of the draughtsman.[1]

Technological change appears to have contributed to the growth in the numbers of technicians both absolutely and relatively. But many technician jobs have been created out of the organizational needs for control and liason between production and administration. Thus many supervisory, quality-control and maintenance functions have gained a need for an oral competence in technical interpretation, as well as an increased facility in the use of instrumentation and electronic control equipment.

As has already been remarked, much of the recent growth represents the 'catching up' taking place within more technically backward industries. The demand appears to be spreading from the nucleus of industries that were early leaders in the employ-ment of technicians. It is difficult to obtain accurate informa-tion on the early part of the century, but what little exists suggests that engineering and shipbuilding were early leaders, with metal manufacturing, iron and steel, and later chemicals making up much of the subsequent growth.[2] Problems which are common to most industries entering the technician labour market were experienced, at least to a degree, by these industries at a much earlier stage in their industrial history. Conceivably they might, therefore, have developed their institutional arrangements for recruiting, training and motivating their tech-nician labour force to a greater extent than would be found in

---

[1] For further elaboration of this point see Chapter 3 and Chapter 5.
[2] For other sources see J. E. Mortimer, *The History of the Association of Engineering and Shipbuilding Draughtsmen*, (The Association, London 1960), and *Manpower* (Political and Economic Planning, London 1951).

other sectors. This study is in fact focused upon these latter
four industries, in which technicians now make up a higher
proportion of the total labour force than in any other industries.
If there are lessons to be learnt they should be demonstrated
within these industries.

Attempts to adjust supply to the growth in demand for technicians

Estimates of the balance of supply and demand for scientific
manpower were made by the Committee on Manpower Resources
for Science and Technology in a triennial survey of engineers,
technologists and scientists, with the 'objective of obtaining the
maximum evidence, both in education and employment, of the
supply and distribution of qualified engineering, technological
and scientific manpower'.[1] The 1965 survey, for the first time,
included information on technical supporting staff.

The survey revealed a total of 622,000 technical supporting
staff, of whom nearly 400,000 were in manufacturing industry.
Another 72,000 were in the public sector of industry, 61,000
in central government and 21,000 in local government—including
technicians employed in maintained schools and further education establishments. Hence the State, in its diverse institutional
forms, is the largest single employer of technicians.

Over all only 4 per cent of technicians had graduate or professional qualifications, and there had been very little change in the
proportion of highly qualified staff over the previous five years,
when compared with figures collected in the chemical and
engineering industries in 1960,[2] though more lowly qualifieid
technicians had more than doubled as a proportion of the total.
Within the pattern of employment public employees stood out as
the best-qualified technicians. Approximately 1 in 15 technicians
working for the Central Government or 1 in 3 working for the
Atomic Energy Authority were professionally qualified. This
ratio falls to 1 in 21 for Local Authorities, but even this is above
the average for private industry and public corporations of
around 1:33. Within this average the span extends from 1:40

[1] Committee on Manpower Resources for Science and Technology, *Report on
the 1965 Triennial Manpower Survey of Engineers, Technologists and Scientists and
Technical Supporting Staff*, Cmnd 3103 (H.M.S.O., London, October 1966).
[2] Survey of Employment of Technicians in the Chemical and Engineering
Industries, *Ministry of Labour Gazette*, December 1960.

Table 1.6   Qualifications held by technical supporting staff, shown as percentage of total technicians

	Degree or equivalent, or Higher National Certificate or Diploma	Ordinary National Certificate or Diploma	Technicians' certificate of C.G.L.I. and equivalent	Other qualifications, training within firms, and no formal training or qualifications
	%	%	%	%
1960 Survey of the Chemical and Engineering Industries	14	13	4	69
1965 Manpower Survey (all sectors covered)	17	14	9	60

SOURCE: 'Survey of Employment of Technicians in the Chemical and Engineering Industries', *Ministry of Labour Gazette*, December 1960; and 1965, *Triennial Manpower Survey of Engineers, Technologists, Scientists and Technical Supporting Staff*, Cmnd 3103 (H.M.S.O., London 1966).

in textiles and also in motor vehicles to 1:13 in aircraft. The ratio within mineral-oil processing is about 1:30. There is little in these results to suggest that formal-training needs at this level are wholly determined by technology. The thesis that process or continuous-flow techniques of production demand higher skills and therefore longer training than those required in mass or large-batch production[1] does not appear to be historically true of technicians. Instead, the most intensively trained technicians appear to be employed in those industries and sectors which rely most on central government finance. The conclusion deduced from national data was substantiated in the plant-level studies that form the basis for this report.

The Triennial Survey of 1965 went on to estimate an increase of 14 per cent in the employment of technical supporting staff over the three years up to 1968, as against an increase of 24 per

[1] See, for example, J. R. Bright, 'Does Automation Raise Skill Requirements?', *Harvard Business Review*, Vol. 36 (1958) pp. 85–98.

cent in the numbers of scientists, technologists and engineers. The Manpower Research Unit's study of manpower trends in the Metal Industries undertaken in 1965 also predicted that the rate of increase in demand for technicians would be substantially slower over the period 1963 to 1968 than over the previous five years. In fact, Table 1.5 shows the numbers of technical supporting staff to have increased by 20 per cent (38 per cent if draughtsmen are excluded) over the four years preceding 1968. The increase in scientists, technologists and engineers amounted to 28 per cent. Over the same three-year period the forecasts made by the Triennial Survey appear to have achieved a remarkable degree of accuracy compared with those of previous surveys. It may have been that the initial survey created a 'self-fulfilling' motive in the employer's response to subsequent surveys. On the other hand, much of the cut-back in the growth of the technician labour force was subsequently attributed to the 1966–68 financial crisis, which did not enter into the Committee's considerations.

The problem of accurately forecasting even short-term needs at national level is made difficult by the lack of manpower planning at company and planning level. In a survey of 262 large and medium-sized firms[1] taken in 1963 the Manpower Research Unit of the Department of Employment and Productivity discovered fewer than one in forty to have undertaken any long-term estimates of their needs in relation to all grades of employees. If it is difficult or impossible for employers to forecast their manpower needs, then it is concomitantly difficult for training and educational institutions to gear their own output to the needs of the market. The more insular and independent employers are, the harder it is to articulate the educational system with that of industry. Much of the supposed individualism of British employers may stem from an inability to forecast their organizational needs and market situation: it is likely, however, that this diagnosis is too simple. Such short-term flexibility in planning as was displayed in the period 1963 to 1968 could not have been achieved without readily available resources of relatively lowly qualified manpower, who could be directed away from technicians' functions to some other pursuit if need arose, or of course vice versa.

[1] Ministry of Labour, *The Metal Industries*, op. cit., p. 10.

The apprenticeship system

The Triennial Manpower Survey extended its concern to the need to fill the gap between supply and demand for technicians which it had forecast. It said that 'in the past' the supply of technical supporting staff had come from the shop floor and from people who had undertaken part-time study. The report doubted whether this would continue to be an appropriate source of supply because more and more children were staying on at school after the minimum leaving age. The report concluded:

... We cannot say that the evidence on supply and demand indicates at present a serious imbalance; but we believe that, because of the growing preference for continuing with formal education after leaving school, the next few years may see the development of an acute shortage unless larger numbers qualify in future.[1]

What is meant by the Report's reference to the future need for formal qualification is left undefined, but there are indications that employers' needs are being influenced from the supply side of the labour market. The Feilden Committee commented:

... Such complaints as the Committee heard from industrialists about the quality of craft apprentices are not a reflection on the quality of the younger generation but a result of the success of the educational system. The clock cannot be set back; old-fashioned apprentice schemes cannot now be expected to produce engineers of quality; arrangements must be made to recruit older apprentices from secondary schools, colleges and universities and to give them appropriate training.[2]

The type of training thought appropriate by Feilden has subsequently gained widespread acceptance in full-time courses in which the syllabus lays great emphasis on the vocational needs of the student. The courses have to be distinguished from the four- or six-year part-time courses formerly taken by most apprentices on a day-release and evening-class basis over the period of a craft apprenticeship. Both courses were, and in modified form still are, taken at technical colleges and are directed at examinations set by the Joint Committees for National Certificates and Diplomas and the City and Guilds of London Institute.

The apprenticeship system stems from a tradition of craft administration which pre-dates most of the industries in which

---

[1] Committee on Manpower Resources, op. cit., para 65, p. 25.
[2] *Engineering Design*, op. cit. (H.M.S.O., London 1963).

it is still practised. It has most usually consisted of a five-year
period of 'learning-by-doing' over a wide range of skills and
associated with a craft identity. Since the days of medieval craft
guilds the apprenticeship requirement has been used by bodies of
craftsmen as a means of retaining the esoteric nature of their
knowledge and passing it on to strictly limited numbers. However, as a 'restrictive or defensive practice', its effectiveness has
been limited to certain sections of a few industries. Employers
have used their bargaining strength to resist attempts at autonomous job-control,[1] but, more importantly, rapid innovation in
organization and technology has enabled them in many instances
to avoid the need to use apprentice-trained labour. Indeed, the
relatively slow growth of training through apprenticeship,
relative that is to the growth in the demand for technically
trained staff, has forced employers into new divisions of labour
or in the employment of 'dilutees'.[2]

Employers have, therefore, been able to secure many of the
advantages of the apprenticeship system without its major disadvantage. The relative cheapness of 'learning-by-doing' and
the return from lowly-paid apprentice labour are still seen as
advantages by some small employers. In general, however, this
was not the case within the plants covered in this study. The
impression gained was that the advantages they saw in apprenticeship lay elsewhere: firstly, in the flexibility gained in plant-
level manpower planning, in particular in career planning, by
this multi-use training. Secondly, it was recognized that although
wastage rates at the end of the apprenticeship training were
high, this could be regarded as part of the cost, which was offset against the gains from recruiting newly qualified ex-apprentices from other firms. In other words it was expected that the
journeyman's 'bag of skills' might include additional benefits
from the knowledge he brought with him from other firms
and other industrial processes.

Against this apparent advantage to the employer the real
costs to the individual worker and to the community are probably

[1] Sidney and Beatrice Webb, *Industrial Democracy* (Longmans, London, 1919
edition) pp. 453–481 and pp. 704–714. For a modern organizational treatment
see A. L. Stinchcombe, 'Bureaucratic and Craft Administration of Production',
Administrative Science Quarterly, Vol. 4, 1959–60, pp. 168–87.
[2] The Report of the Committee on Technician Courses and Examinations
gave it as its opinion in 1969 that '. . .probably still the majority of the country's
body of technicians have no relevant educational qualification at all'.

high. The wastage rate is enormous in the early years of the extended evening and day-release courses at local technical colleges. In one study only 6.7 per cent of the students entering O.N.C. courses in four colleges completed the course in the minimum time.[1] The methods of on-the-job training are also notoriously inefficient. Yet it appears from all of the numerous reports on the training and education of technical staff that the apprenticeship system has earned an established position in supplying new entrants to the technical labour market that will be difficult to dislodge. Professor Peck suggests that 'the preferences of some British firms for on-the-job apprentices may also have discouraged the expansion of the supply of engineers.'[2] As late as 1964 over half of British engineers were trained outside universities, compared to 25 per cent of American engineers, on the other hand 90 per cent of British pure scientists came out of universities. The attachment of British employers to part-time education at local technical colleges has been seen as a major constraint on the expansion of professionally qualified engineers.[3]

Before entirely accepting the economists' judgement on apprenticeship, one ought perhaps to look at less manifest social functions it may fulfil. In particular one of the more important social functions of apprenticeship was what has been termed a 'cooling-out system.'[4] By this is meant the adjustment of the aspirations of ambitious young recruits to a realistic level of achievement within the labour market or, from the employer's point of view, within the limits of what the firm could provide. Young recruits would normally go together through the same initiation course within the company or plant before being allocated to various departments to do their 'on-the-job' training in workshop techniques. At the same time they began a parallel course of part-time study at a local technical college. Here there was, and is, some streaming into the more academic National

[1] P. F. R. Venables, *Technical Education*, Bell, London 1956, p. 239.
[2] Richard E. Caves *et al.*, *Britain's Economic Prospects* (Allen and Unwin, London 1968).
[3] In particular see *15–18*, *A Report of Central Advisory Council for Education (England)*, under Chairmanship of Sir Jeffrey Crowther, Ministry of Education, 2 vols. (H.M.S.O., London 1959), and *Half Our Future*, Central Advisory Council for Education, Department of Education and Science, (H.M.S.O., London 1963).
[4] For a more comprehensive description of this process see Burton R. Clark 'The "Cooling-out" Function in Higher Education', *American Journal of Sociology*, May 1960, pp. 569–76.

courses and the less abstract curriculum required for the City
and Guilds examinations. There is some flexibility in movement
between courses, but the greatest degree of compensation allowed
to the academic 'under-achiever' is in the work context. For it is
here that the apprentice can, if he wishes, become identified
with a craft group and take his behavioural norms from them.

Usually a management decision is taken in the penultimate
year of the apprenticeship to place the apprentice in the work-
situation which will determine his career opportunities. By that
stage the academic achiever is already marked out for the
laboratory, the drawing office, or one of the design functions.
For those who fail to reach these or similarly prestigious positions
there is a whole range of secondary goals in respect of which even
that of 'craftsman' represents a successful outcome of this
arduous and time-consuming training.

The very uncertainty of the goals set for the apprentice technician makes the lack of institutional means to their achievement
less of an immediate problem. Even after the first glow of
achievement has worn off, the ambiguity of his status does not
allow him to make a very firm assessment of his career prospects.
Nor does he feel any immediate desire to condemn a training
and an education which has set him apart from his less successful
former colleagues, and which now constitutes his claim to separate treatment in the work-place. Unlike his contemporary in the
university, he appears much less likely to want to change 'the
system'; rather he may wish for more recognition from those in
authority, such as increased financial rewards for his hard-
earned position in the social hierarchy.

The shift to college education

The national need for more trained technologists and technicians
has led to a series of government reports and policies designed
to (i) encourage the development of technological degree
courses; (ii) extend existing technical certificate courses in colleges;
and (iii) make further provision for training within industry.
All these methods, while increasing numbers, have had the
effect of formalizing qualifications, clarifying career opportunities, and resolving ambiguities of status for the technician.

The results appear potentially frustrating for those who
entered their jobs with aspirations created by a different and

more informal task-centred system of qualifications and rewards. For those who have yet to decide on an occupation their response to the proposed new norms may not accord with the values that lie behind them.

(1) *Technological degrees*

The development of technological degree courses has been somewhat confused by the debate over whether this should take place in the universities or technical colleges. The first post-war Government Committee report was most explicit.

> ... Industry must look mainly to Universities for the training of scientists, both for research and development and of teachers of science; it must look mainly to Technical Colleges for technical assistants and craftsmen. But both Universities and Colleges must share the responsibility for educating the future senior administrators and technically qualified managers of industry...[1]

This dual development has in fact taken place. On the one hand, the attempt to provide technologists with a place in the university system as firm as that enjoyed by pure scientists was highlighted by Sir Winston Churchill's famous Woodford speech in 1955.[2] This served as the inspiration for the foundation in 1960 of a new college within the University of Cambridge dedicated to technological studies, and it has been followed by the formation of new technological universities at Loughborough, Manchester, Brunel and Surrey. On the other hand, the Percy Committee recommendations and those of later Government Committees[3] that selection of a strictly limited number of existing Technical Colleges be allowed to develop technological courses of degree standard, meant that from 1950 onwards the major technical colleges began to gain increasing support for their claim to be consulted on and to participate in the education of technologists. They were rewarded in 1956 by the announce-

[1] Percy Committee Report, *Higher Technological Education* (H.M.S.O., London 1945), p. 6.

[2] In this speech Churchill appeared greatly influenced by yet another emulative comparison which appeared in N. de Witt, *Soviet Professional Manpower*. (See an article in *The Times Educational Supplement*, December 30th, 1955: 'Scientists and Technologists in U.S.S.R.', by Prof. S. Zuckerman.)

[3] See National Advisory Council on Education for Industry and Commerce, *The Future Development of Higher Technological Education* Ministry of Education (H.M.S.O., London) p. 9.

B

ment of courses to be run within a limited number of recognized colleges leading to a degree-level Diploma in Technology.[1] This was followed in 1965 by the initiation of degrees awarded under the Council for National Academic Awards, a proposal of the Prime Minister's Committee on Higher Education under Lord Robbins (1961–1963). Yet, in the same year, 1965, the Secretary of State for Education and Science reaffirmed that the historical dichotomy within Higher Education between vocationally based technical colleges and academically oriented universities would be regarded as a guiding principle for future development.

Since that time the rate of expansion in Higher Education has shifted rapidly towards the publicly controlled (technical-college) sector. This shift has been reflected in the growing success achieved by technical-college courses, as shown in the table below.

Table 1.7 Students in full time advanced further education by type of course, England and Wales

Type of Course	Thousands 1961–2	1967–8	Percentage growth
London degrees	4.2	13.7	226
C.N.A.A. degrees and Dip. Tech.	1.3	10.1	677
Higher National Diploma	4.0	14.2	255
National art qualifications	5.8	6.7	16
Professional qualifications, etc.	8.5	21.3	151
Total	23.8	66.0	177

SOURCE: Richard Layard, John King and Claus Moser, *The Impact of Robbins* (Penguin Education Special, London 1969).

The development of technological degree courses in the universities and technical colleges has had three main effects on potential technicians:

[1] *Memorandum on Higher Technological Education* (H.M.S.O., London 1956). The Memorandum and the White Paper that preceded it (Cmnd. 1254 *Better Opportunities in Technical Education* (H.M.S.O., London 1961)), made reference to new sources of public finance for these colleges, other than those which already existed in the Local Authorities. The question of allocation of scarce resources within Higher Education was, and remains, a main theme of the binary controversy.

(a) Many more students are taking A levels at technical colleges, perhaps as a perceived preliminary to a degree course. The Robbins Committee predicted that in 1967 there would be 3,800 home students obtaining two A levels in colleges of further education. The actual output was 7,700. This was partly attributable to a passing bulge in this age cohort; the proportion of any single age-group staying on until 17 or 18 at technical college appears to have remained relatively stable. But within the numbers of adolescents staying on there is an increasing trend towards taking A level G.C.E.[1] This contrasts with the trend in entries to National Examinations, which actually declined after 1964: in the case of the Ordinary National Certificate or Diploma, both of which are crudely comparable to G.C.E. A levels, the fall was from 40,420 in 1964 to 22,731 in 1968.[2]

While A level students make up only about 10 per cent of their age cohorts in technical colleges, and undergraduates are also only a small proportion of the whole, the increasing importance of degree-oriented courses is a constant reminder to part-time students serving apprenticeships, as to their employers, that an alternative method of entry into the ranks of technologists is open to them. Moreover, this entrance is not, like the university, far beyond any reasonable hope or expectation that they might have entertained in their secondary school.

(b) Large numbers identify with technologists and aspire to careers as technologists. These technicians would wish to take a degree if they could, even if for large numbers of them who left a secondary modern or secondary technical school at the age of 15 or 16 the chances of taking a C.N.A.A. degree or even A levels are still fairly remote by any objective measure. In a study of technical-college students carried out by Professor S. F. Cotgrove in 1956–57 it was found that over half of part-time students taking courses for a National Certificate or Diploma would have preferred to have been taking a degree. Of those who did not wish to take a degree

[1] R. Layard et al., op. cit.
[2] Report of the Committee on Technician Courses and Examinations (Dr H. L. Haslegrave), National Advisory Council on Education for Industry and Commerce, Department of Education and Science (H.M.S.O., London 1969), Table 2, pp. 13–14.

the majority mentioned 'ease of entry' as a reason for entering their present course.

Differences between the entries to courses seemed to be much more closely associated with previous secondary education than with any alternative external factor other than the influence of the employer.

Table 1.8 Previous secondary education received by sample of students attending evening courses in industrial and scientific subjects at two polytechnics

Secondary Education	Univ. Degree	G.C.E.	O.N.C.	H.N.C.	Prof.	C. and G.
	%	%	%	%	%	%
Modern	12.5	18.6	40.5	33.6	22.5	88.4
Technical	10.4	8.5	20.2	28.1	14.1	2.9
Grammar	58.3	57.6	22.3	25.8	40.9	5.8
Public Day/Bdg.	16.7	10.2	10.6	8.9	16.9	2.9
Others	2.1	5.1	6.4	3.6	5.6	—
Total	100.0	100.0	100.0	100.0	100.0	100.0

SOURCE: S. F. Cotgrove, *Technical Education and Social Change*, Table 10 (Allen and Unwin, London 1958).

In this desire to become graduates, part-time technical-school students are of course displaying similar values to those represented in the national order of priorities which have placed a premium on an academic approach for the teaching of technology. They are values which have served to differentiate technicians from shop-floor workers, and at various times have caused them to identify themselves with the scientific community. Increased opportunity for those with an appropriate level of educational attainment to avail themselves of degree courses simply serves to emphasize the relative deprivation felt by those who do not have the appropriate prior qualifications.

(c) The third and probably most important effect of the growing number of graduate technologists on the technician's status and market situation is on his 'promotion ladder'. As a result of the substitution of external sources of qualification for aptitude and competence in the job as criteria of ability the career chances of technicians may well be considerably foreshortened. So long as over half of British technologists are apprentice-trained it may still appear possible to the ex-apprentice that his indentures are the managerial baton that he carries in his knapsack, however prolonged the route to the top. Once, however, the means to success becomes more formalized, then a whole range of intermediate steps in the form of academic successes become apparent. New comparisons made outside the work situation also become more important. Of course, such formal achievements have always been present in the award of National and City and Guilds Certificates; but the immediate financial incentives and other rewards given by employers for passing these examinations have generally been small. The system has been, and many managers maintain is still, one that recognizes ability at the job rather than external qualifications. So long as this was so, a general equilibrium could be maintained within the growing numbers of technicians recruited from a range of diverse sources, including many shop-floor workers who showed an aptitude for the job. Once managers began to look to external sources of qualification these more localized criteria of competence came under increasing strain.

(2) *Technical certificates*

The effect of government policy towards technical colleges has also been in the direction of a more clear-cut system for obtaining qualifications. The primary objective of the 1961 White Paper entitled *Better Opportunities in Technical Education* and of some of the later work of the Industrial Training Boards set up after the passing of the Industrial Training Act in 1964 was to bring about a major restructuring of the curricula of the National and City and Guilds courses being taught in technical colleges. The work of the Boards and the more recent proposals of the Haslegrave Committee in 1969[1] are designed to carry through this concern to the point where they have a major influence in reshaping plant training schemes. Both have as primary objectives to discover, define and satisfy 'the needs of industry' in regard to training.

[1] *Report of the Committee on Technician Courses and Examinations*, Department of Education and Science (H.M.S.O., London 1969).

The effect of the 1961 White Paper was formally to recognize that streaming already existed within the education that accompanied the apprenticeship system.[1] This it did by suggesting an introductory course over the first one or two years of college attendance. There the young trainee would be assessed for entrance to the more academically orientated Ordinary National Certificate or Ordinary National Diploma course, or for one of the newly designed Technician Courses provided by the City and Guilds of London Institute for the more 'practically oriented'. These courses, pursued through day-release or block-release facilities afforded by employers, would last for two or more years, with the possibility of further progression to Part III of a Technicians' Course, or, for those with a better grasp of theory, to Higher National Certificate or Higher National Diploma over another period of two or more years. Entry standards for National courses were also to be tightened, but greater movement allowed within the confining elements of courses available and employer preferences. These measures were designed to reduce 'wastage'.

It has already been mentioned that the annual rate of entrants to O.N.C. and O.N.D. courses fell by 74 per cent over the period 1964 to 1968 when the rate was rising to A level and other courses. Over the same period annual entries to H.N.C. courses rose slightly by 6 per cent from 18,440 to 19,551, reflecting an increase in the numbers who had previously taken O.N.C. and O.N.D. courses before the subsequent slump. But the really impressive increases all took place within full-time or block-release courses. This latter type of education was recommended by the Crowther Committee as the only satisfactory means of bringing students and teachers together for long periods of time. Numbers entering these courses annually increased from 24,493 to 45,853 over the four-year period. However, a large proportion of these were taking the *practically oriented* courses of the C.G.L.I.: numbers of trainees taking such courses on a block-release basis rose from 4,338 in 1961 to 34,746 in 1967.[2]

There are a number of pressures at work to bring about this

[1] The majority of such training schemes in existence then (and now) are not apprenticeships in the sense that recruits sign binding indentures to complete a stated period of employment.

[2] All figures in this paragraph are extracted or computed from tables produced in the *Report of the Committee on Technician Courses and Examinations*, Department of Education and Science (H.M.S.O. London 1969).

change to practically based training. There have been many outspoken critics of National Certificate curricula, and the movement towards more task-focused courses has supporters among planners, teachers, and—most important—among management. Financial inducements offered by the Training Boards have made the arguments of educationalists on the wisdom of the full-time 'block-release' method of training appear more attractive than hitherto, and hence have facilitated the increase in full-time courses. However, the choice of young people who are 'voting with their feet' by entering full-time academic courses has indubitably helped to create the new norm.

Within the new full-time technical-school population the lines of demarcation are therefore becoming increasingly clear. If one can extrapolate from the findings of Professor Cotgrove's small sample, the aspirations of the young people doing practical courses may not necessarily adjust to levels defined by their teachers and employers. They may, for example, look to other qualifying bodies such as professional associations to provide them with the symbolic prestige that higher education has refused them.

Yet the preferences of both employers and national planners appear to be pushing vocational selection further and further back into the educational system and attempting to gear post-C.S.E. or G.C.E. training to these specific ends. Thus the Committee on Technician Courses and Examinations which reported in 1969 suggested:

> The extra year of compulsory secondary education, and the further development of C.S.E. examining techniques leading to better prognosis of young people's true capabilities for undertaking further education and training will undoubtedly reduce the need for the General (introductory) type of course, and indeed we think it should be possible to dispense with it altogether when the school leaving age goes up.

Instead, it recommended that for school leavers of 16 there should be a rudimentary two-year technician's course, which should embody the diagnostic function of the General Course and allow transfer to a craft route. The attainment of a Technician's Certificate after two years should allow the able and aspiring trainee to continue to a Higher Technician's Certificate or Diploma. The Report argues quite fiercely (and perhaps a little desperately?) that a degree is not the equivalent of the H.N.D. and that the retention of an examined course similar to this latter is necessary, if only as a suitable 'fall-back' course for failed undergraduates.[1]

The proposed Diploma can in fact be seen as a step on the ladder to a degree in the proposed scheme. In either capacity it is very much a second-class option academically, which may be avoided by qualifying for a full-time degree course through A levels and thus avoiding the hazards of part-time vocational education. The success of the proposed new qualifications appears to depend very much on their acceptance by management and by the technicians themselves. The fate of the short-lived Diploma in Technology, awarded by the C.N.A.A. during its first years but abandoned because of its failure to acquire the prestige of a degree, is perhaps a measure of the difficulty of this task.

(3) *Training in industry*

Even here there are proposals for greater distinctions as between craft and technician training, although there is some employer-resistance to this.

The Industrial Training Act of 1964 had as its first aims 'to ensure an adequate supply of well-trained men and women at all levels of industry and commerce, and to secure an improvement in the quality and efficiency of training'. The method used was to focus the effect of its activities on plant-based training, though this was normally seen as complementing changes in further education. The Central Training Council set up under the Acts has brought into being a range of industry-based Boards financed by *per capita* levies on all but the smallest of employers within the appropriate industry. Those who satisfy their Board's standards in the training they provide for the employees upon whom they pay a levy will receive a grant to offset the cost of this training. Hence the costs of training are spread in such a way as to set a premium on the provision of an 'approved' scheme. Most observers see the effect of the Act as being beneficial in terms of improving the 'overall awareness' of training needs among employers. One of the more serious criticisms is, however, that the Boards so set up have seen it as necessary

[1] op. cit., paras. 224–226

to define the purposes of their training within fairly restricted occupational boundaries. Thus one of the early dichotomies made by most boards was that between craftsman and technician; from this has stemmed two quite separate training schemes, linked only by so-called 'bridging courses'.

In 1969 the Engineering Industry Training Board went further and produced a report in which two distinct levels of job-clusters were identified within the technician category. From among the general classification of 'other technician' they separated out an élite of 'technician engineers', whom the Board maintained required special abilities in the use and communication of information, and whose skills encompassed those of abstract understanding and the ability to organize and give direction to the work of others. The creation of this dichotomy has widespread implications.

More immediately it has brought about a requirement that employers should adopt a three-year training programme for 'technician engineers'. It is suggested that the curriculum will become the model to which the Board will refer in the allocation of training grants after September 1971. The content of the suggestions is important, not only because of the size of the Engineering Industry, but also because they contain principles which appear in discussions on technician training in other industries and which will provide an important precedent.

They include the suggestion that companies should provide up to a year of General Training within the company upon such subjects as design appreciation, manufacturing practice, commercial matters, control techniques and communications. This type of training would make the higher-grade technician better equipped to adopt a managerial career, and indeed the Board recommend that such an élite should require complementary further education to a level within the range of the Full Technological Certificate of the C.G.L.I. and the Higher National Certificate or Diploma in engineering subjects (para 10, Introduction).

The effect of such a distinction is to emphasize the selection process that should already (since 1961) take place between craft and technician training in the first year of training. When these proposals are adopted it will be extremely difficult for a member of the 'other technician' grades to overcome training differences in both the place of work *and* the technical college. One observer has commented, 'The E.I.T.B's booklet on training for technician engineers concentrates exclusively on the training needs of school leavers: it makes no reference to the possibilities of upgrading for experienced craftsmen.[1]

However, the major sources of opposition to the proposals appear at the moment to be coming from the employers rather than from employee representatives. For many training officers the notion of offering a general training going beyond the specific task requirements of the firm appears to be quite alien. To Feilden[2] as to Hazlebury it seemed evident that such task-specific training would not satisfy a perceived need for technically trained middle management, which has up to the present been drawn from the ranks of apprentice-trained technicians. Hazlebury makes this function of technician training extremely explicit:

> ... General studies can play a particularly valuable part in rein-
> forcing the efforts made during the vocational studies to develop
> in the future technician the communication skills which are to be so
> important to him in his future career. They should also be of real
> benefit in preparing him for the supervisory role he will almost
> certainly be called upon to undertake in the course of his career.[3]

Yet the chances of promotion for the individual technician, while varying very greatly within the job context, have always been relatively small. Since the ratio of technicians to technologists has been increasing throughout the post-war period, the objective possibility of promotion to administrative rank has become progressively smaller. The new position in which the numbers of technologists have been increasing rapidly over the past few years has, far from relieving the situation, made it manifestly worse. Manifest, that is, to the technician himself: for the new men have generally been graduates, a fact which does not appear to have escaped the growing numbers of young people exercising their preference for A level G.C.E. rather than City and Guilds or O.N.C. courses. One has only to refer back to the recent history of technical education to see how difficult it will be to establish a Full Technological Certificate (as is suggested for the technician engineer), which is as prestigiously attractive as

[1] Department of Employment and Productivity, *Report on Technician Training* (H.M.S.O., London 1970).
[2] *Engineering Design*, op. cit. (H.M.S.O., London 1963).
[3] *Report of the Committee on Technician Courses and Examinations*, para. 270 (H.M.S.O., London 1969).

a university degree. The argument must have been clinched for many students by the growing acceptance by management themselves of external academic qualifications. This was a prior condition to the high market value acquired by professional engineers during the 1960s.

For the older technician or for the many thousands whose secondary education is such that even a C.N.A.A. course is out of the question, the chances are that the promotion ladders will not only be shortened, but will be seen to be shortened. Trainees are increasingly being streamed at 15 or 16+, and learning, both inside as well as outside the work situation, will reinforce this distinction. A number of observers have suggested that the general 'in-plant' training towards which the employers take exception is indeed unnecessary and that more specific training might be 'packaged' into modules to take place over six to twelve months, which in various combinations would make up a technician apprenticeship.[1] This technique might hold many advantages for employers but without some universally acceptable general training, it seems possible that the craftsman will lose his 'bag of tools' entirely.

The provisions made by Industrial Training Boards may therefore go some way towards remedying the short-term market deficiencies in technician skills. But in doing so they are helping to bring definition to a formerly ambiguous occupational situation. Like the introduction of C.N.A.A., they tend to take individual success indicators out of the work-place and into the classroom. The ambitious young technician will know in a relatively clear way and at a relatively early age whether or not he will be able to emulate his supervisor, chief draughtsman or laboratory head. To educationalists and those concerned with the supply of trained technicians it has often seemed that the craft apprenticeship has resulted in an 'over-recruitment' of able youngsters, many of whom were capable of far more than the toolmakers' or electricians' tasks to which they eventually graduated. However wasteful this may have been for the economy as a whole, the individual employer has never had to bear such social costs, and, for the reasons already given, the private benefits in terms of flexible manpower planning and career development were considerable. The overwhelming major-

---

[1] This approach to training is already widely accepted in craft apprenticeships, which are, of course, often the prelude to a career as a technician.

ity of managers interviewed in the course of this study were ex-apprentices who saw no reason why the 'leadership potential' of the present generation of technicians should not 'emerge' in the same way as had their own. This attachment to a personal prototype is familiar to all students of social change. It may well account for current disjunctures in the market for technologists as well as that for technicians.

Conclusions

There are many views of the problems involved in matching supply to demand. For the educationalist and manpower planner the problem is resolved in the creation of a range of appropriate formal qualifications matched as far as is possible (with the co-operation of management) to the diverse range of skills required of the technician as he progresses through his career. A second view is represented in the generally shorter-term attitude of management, who for the most part prefer to handle technician recruitment and career development on an individual company basis, yet who appear to pay increasing attention to external qualifications in their current recruitment.

There is a third view, that of the technician himself: this latter provides the major focus for this study. The more clearly the technician is located within the occupational spectrum by other institutions—educational and managerial—the more clearly is he able to match this placement against his own 'established expectation',[1] originally created in the induction to his career provided by management itself. His identity with the technologist or manager has rarely been overtly challenged, and his proclivity to adopt the style and nomenclature of these higher-status groups have been accurate guides to the direction of his ambition. The ambiguity of his role and status, which causes Government Commissions to talk of 'scientists, technologists and supporting staff' also enables the leaders of their trade unions and professional associations to claim parity of consideration for their problems. The possible dissonance created by an emerging status-inconsistency is one that could lead (and may well have led) to widespread feelings of deprivation, and consequently to inter-group conflict.

Whatever the form and process of the industrial relationships

[1] Sidney and Beatrice Webb, *Industrial Democracy*, p. 68

emerging from this complex and dynamic situation, it is certain that the informal and even highly personal form in which they have evolved within the work situation is to be increasingly formalized and invaded by external constraints. It seems inevitable that with a technician labour force making up a substantial and increasing proportion of the total labour force, those concerned with planning the economy should seek to shape the market for technicians. The problems of bringing about these changes go far beyond those of drafting procedural changes. The purpose of this study is to reveal the attitudes of those who are being subjected to occupational planning, from their education to their induction and career development, on a scale never before experienced by any single emerging group in history. It will describe something of their work situation and their market position, and will suggest some of the unanticipated effects of the attempts to shape the dynamics of occupational change.

2 The Definition of a Technician

One of the themes of the last chapter was the lack of definition given to the occupational boundaries of technicians. This difficulty has its repercussions even at the most basic level of investigation. Available statistics on technical staff are confusing because data from one source do not necessarily match up with those from another. For example, the Report on the 1965 Triennial Manpower Survey quoted earlier gave a figure for 'technical supporting staff' in manufacturing which, at 398,374, was no less than 50 per cent greater than that given in the Survey of Occupations of Employees in Manufacturing Industry taken in the same year by the Ministry of Labour (261,130). Although the difference in survey techniques employed in these studies accounts for some of this discrepancy, it seems likely that so great a disparity stems from a difference in the jobs included in the population.

The failure to reach a consensus on a suitable definition is in part due to the sheer novelty of the situation. As late as 1949, the term 'engineering technician' had not found its way into the American *Dictionary of Occupational Titles*,[1] and until the 1960s many British manpower forecasts simply aggregated 'scientists, technologists and technicians' in a manner which provided little guide to the actual situation. Yet with the wisdom of hindsight it was possible for one Government committee to see that:

[1] See W. M. Evan, 'On the Margin: The Engineering Technician', in P. Berger (ed.), *The Human Shape of Work*, (Macmillan, New York, 1964).

This is not merely a matter of new occupations being created although this is an important factor in the situation in all industries. The jobs now being identified and designated as technician jobs have been done in these industries for a long time past; it is only now that the industries are beginning to accept that the people doing them are technicians, and to think of them and describe them as such.[1]

To a certain extent the emergence of a new occupational group can only be recognized *ex post facto:* after it has established an institutional identity of its own or had one created for it by other existing interests in society. A new occupational group inevitably takes time to acquire a defined character which is understood and accepted by those in the occupation, by the employers using the occupation, and by the government, which is ultimately responsible for establishing the public status of the occupation by including it in its official statistics. Differences of opinion and practice between those following the occupation, their employers and the government, may, as in the case of technicians, create considerable confusion and give rise to a serious and continuing problem of definition.

The historical approach of managers to technicians has been characterized by a desire to maintain as flexible a labour supply in these grades as possible. The idiosyncratic needs of managements have been determined by their perception of production and research and development requirements within their particular establishments. In general there has been little attempt at manpower planning or long-term career development for technicians. The great diversity of job titles discovered in all surveys, including the present one, demonstrates a highly individual approach to the grading of technicians as well as the heterogeneous nature of the tasks at which they are employed. There is, moreover, a sizeable degree of overlap in terms of the performance of common tasks with those roles which are more precisely defined as those of technologists, or, at the other end of the spectrum, with those of craftsmen, and, as even the present study revealed, with those of bench-trained manual workers. As has been demonstrated in the first chapter, technological constraints have allowed wide variations in the qualifications required to

[1] *Report of the Committee on Technician Courses and Examinations* (Dr H. L. Haslegrave), Department of Education and Science (H.M.S.O., London 1969). p. 5, para. 8.

perform even the most necessary tasks.[1] From an initial reliance on apprentice-trained craftsmen there appears to have been a tendency to include a broader spectrum of untrained recruits into the labour pool from which technicians could be drawn. By 1969 the Engineering Training Board found that in a wide survey of technicians undertaken in the industries covered by the Board, 30 per cent had been craft apprentices, 31 per cent had technician training of an apprenticed form, while 25 per cent had no kind of apprenticeship training whatsoever.[2]

The great diversity in tasks and work-situations, and the heterogeneity among the performers, has bedevilled a number of attempts at functional definition made by national and international agencies. One of these reporting on technicians stated:

> They are employed in an immense variety of roles in industry and the public service; in assisting research and development; in draughtsmanship, design, instrumentation and control of various kinds, computer programming, servicing and maintenance, method and work-study, quality control, estimating, rate-fixing, technical writing, inspection, and increasingly too on supervision. It would be almost impossible to give anything like a complete catalogue of technician occupations since new ones are continually emerging and developing.[3]

Not surprisingly, a report of a joint working party set up by the United Kingdom Automation Council and the City and Guilds of London Institute, in October 1966, considered the search for an overall yet precise definition of a technician's role to be 'unnecessary, time-consuming and impossible of achieve-

---

[1] No attempt has been made at detailed job analysis in this study. What slight evidence is available on the determinant effects of socio-technical environment on the task composition of jobs is contained in the *Report on an enquiry into the employment and training of technicians in the engineering industry 1969* (Research Planning and Statistics, Engineering Industry Training Board, March 1970). This survey included an activity sample with log-book recording among 1,044 technicians. Apart from the Research Design and Development function the task composition of the technicians' job seems fairly flexible across a range of industries. It seems therefore that technological constraints on the employers' area of choice in the organization of work under job titles may also be quite wide over most of the functional areas covered by the 'technician' nomenclature.

[2] *Report on an enquiry into the employment and training of technicians in the engineering industry*, op. cit.

[3] *Report of the Committee on Technician Courses and Examinations* (Dr H. L. Haslegrave) Department of Education and Science (H.M.S.O., London 1969). op. cit., p. 7, para. 20.

ment'.[1] Reporting on a three-industry sample investigation in member countries, an O.E.C.D. Report of 1966 commented:

> Certain common factors appeared in the three investigations. The first was a general lack of definition in the function performed. In very few cases did a job actually bear the name of 'technician', and in order to identify the right people for investigation it was often necessary first to reach agreement with the employers or supervisors as to which functions could properly be regarded as technician ones and which people were performing them.[2]

Some national agencies have been concerned with bringing about an appreciation of the importance of technicians and the need to establish their status both in the eyes of management and among technicians themselves. This purpose and the reasons for it were perhaps most explicitly stated in this extract from the Report of the Committee on Technicians' Courses and Examinations 1969:

> In many instances their own employers have yet to recognise the distinctive nature of their jobs; in just as many cases the individuals themselves have yet consciously to realise that they are members of a segment of the country's skilled manpower who need specific attributes of intellect, skill and experience to do their jobs and who also need a training and education programme different from that undertaken by those who work above or below them. But enough is known to recognise that the work technicians do in maintaining the industry and commerce of the country indeed the whole fabric of its society, is of first-rate importance, and that their education, training and status should reflect this.[3]

Yet the primary purpose of most other national planning agencies is expressed in their uniformly high concern to ensure a sufficient supply of appropriately trained and educated technicians for this huge and expanding job market. In doing so they are more generally involved in the allocation of educational and training resources over a range of occupations and of therefore defining technicians in terms of a hierarchy of skills and knowledge. The very reason that called them into existence makes the task of these committees more difficult. An early attempt

---

[1] *UKAC Record*, No. 9, p. 9.

[2] *The Education, Training and Functions of Technicians: United Kingdom* (Directorate for Scientific Affairs, O.E.C.D., Paris 1966).

[3] *Report of the Committee on Technician Courses and Examinations* (Dr H. L. Haslegrave) Department of Education and Science (H.M.S.O., London 1969). p. 8, para. 22.

at a definition was made in a Government White Paper of 1956: 'a technician is qualified by specialist technical education and practical training to work under the general direction of a technologist. Consequently he will require a good knowledge of mathematics and science related to his own speciality.'[1] In more explicit terms, a later report of the Ministry of Labour stated that: 'The qualifications of technicians include Higher National Diploma, Higher National Certificate, Ordinary National Diploma, Ordinary National Certificate, the City and Guilds Technician Certificates and the General Certificate of Education at advanced level . . .'[2]

Yet when the 1965 Report of the Committee on Manpower Resources for Science and Technology[3] looked at the qualifications of technicians, by industry, it discovered that having a qualification was not a *sine qua non* of the technician; 58 per cent of technicians in British industry had no formal technical qualifications, not even those of the City and Guilds Institute. In a small study of 94 technicians carried out by Professor Stephen Cotgrove in 1962, 55 per cent had no technical qualifications.[4] On the basis of these and other studies the Haslegrave Committee concluded that 'there is a tendency in some quarters to identify technicians by reference only to educational courses they have taken. This we are sure is wrong at a time when a very large part—probably still the majority—of the country's technicians have no relevant educational qualification at all.'[5] Instead the committee urged that 'technicians must be identified in the first place by the function they perform'. This must indicate a massive job-analysis exercise, a task which the Committee sees as being properly that of each Industrial Training Board to perform within its own industry. A number of fairly specific studies had in fact already been undertaken, particularly of Automation and Instrument Technicians[6] and of technicians

[1] *Technical Education*, Cmd. 9703 (H.M.S.O., London 1956), p. 2

[2] *White Paper*, Better Opportunities in Technical Education p. 5, Cmnd. 1254 (H.M.S.O., London 1961).

[3] Op. cit.

[4] See S. Cotgrove, 'Technicians in Industry', *Technology*, November and December 1960; 'Does the Boss Care?', *Technology*, August 1962; 'The Relations between Work and Non-Work among Technicians', *Sociological Review*, Vol. 13, No. 2 (New Series) July 1965.

[5] Op. cit., p. 5.

[6] 'Investigation concerning Technicians associated with automated processes', *U.K.A.C. Record*, No. 5, February 1965.

in the Iron and Steel Industry.[1] More recently a number of Industrial Training Boards have published technician training recommendations, which are to be ultimately backed by the sanction of training grants allowed on the basis of an employer's acceptance of the proposed schemes.

A feature of some proposals stemming from Industrial Training Boards has been the division of technician work in a manner which is common in France.[2] French technicians are often categorized as *technicien supérieur,* and are expected to attempt more complex and challenging work than that of the lower grade of *agent technique.* There is evidence of a similar division of labour developing in Holland also. In making this division it is intended that the lower-level technician should relieve his more highly qualified (and possibly more ambitious) colleagues of routine and monotonous tasks, enabling him to utilize his higher ability or training to a greater extent than before.

The most far-reaching of these proposals to be put forward in Britain is that of the Engineering Industry Training Board, on which comment has already been made (see Chapter 1). However, the Iron and Steel Industry Training Board has indicated the desire, which is widespread among the industry's management, to keep the definition of technician as exclusive of the lower-level assistants as possible:

> ... a technician should be competent to apply in a responsible manner proven techniques which are commonly understood by those who are expert in a branch of technology or techniques specifically defined by technologists. He should have practical knowledge and experience to assist him to diagnose problems, work out details of a task or operation, carry out the work himself and in some cases exercise supervisory or advisory duties.[3]

By contrast, the laboratory assistants and other lower-level technical workers generally support technicians and other staff,

---

[1] Most particularly *Report on the Recruitment and Training of Technicians,* April 1961, and *Report on the Education and Training of Technicians in Laboratory and Quality Control Departments,* British Iron and Steel Federation. October 1962. See also *Recommendations on the Training of Technicians and Technologists,* (Iron and Steel Industry Training Board, May 1966).

[2] See P. Berger, *The Human Shape of Work* (Macmillan, New York 1964).

[3] *Recommendations on the Training of Technicians and Technologists,* op. cit. (I.S.I.T.B., 1966). Also *Training of Chemical, Metallurgical and Fuel Laboratory Assistants and other equivalent jobs* (I.S.I.T.B., 1968).

and their work usually consists of 'testing, analysis at an elementary level and of recording the results for others to use'.[1]

This division of duties appears to be reaching for an ideal, rather than representing the reality as it exists either in the iron and steel industry or in engineering. A report on an inquiry carried out by the Engineering Industry Training Board in 1969 found

> ... the kind of activity carried out by the whole range of 'technician engineers' and 'other technicians' is basically very similar, with a wider area of overlap between the two levels than had been expected. For example there is little evidence to indicate that the activities of planning evaluation, control etc., are any more within the scope of the technician engineer than of his counterpart at a lower level. Nor does the former spend substantially more time on supervisory activities. Similarly the technician engineer does not spend very much less time on the relatively simple activities of measuring, testing, inspecting, etc. than does the other technician.[2]

On the other hand this Engineering Training Board survey, which is by far the most sophisticated of its kind to be attempted in British industry, found 'that the differences between some sub-groups of technicians are almost as great as those between technicians in general on the one hand and administrative staff, craftsmen or operators on the other, and that, while certain particular kinds of technician jobs are easy to identify, the whole range of technician jobs is not easy to describe or categorize'.[3]

The clear-cut differences in skills or authority only occurred at the occupational boundaries, and only then in one particular work situation. The activities of many 'other technicians' in Research and Development made them more like craftsmen; they had no technical qualification and spent almost a quarter of their time on assembly and manufacture. Few technicians were engaged in managerial or planning operations, though once more in Research and Development 'technician engineers' were often given more discretion in design and communications.

Each of the three means to define an occupation—whether through formal qualifications or training, the functional definition

---

[1] *Training of Chemical, Metallurgical and Fuel Laboratory Assistants and other equivalent jobs,* op. cit., p. 1.
[2] op. cit., p. 29, para 125.
[3] ibid, p. 30 para 130.

of tasks, or the placement of an occupation in relation to an organizational and market hierarchy of skills, intellect and/or authority—are inherently unspecific and difficult to quantify. The most exact guide is provided by that of task- or job-analysis, yet the tasks that have accrued to the technician are largely those which have evolved in a locally specific manner, without the constraints which might have been provided by some external control at national or industry level. The very heterogeneity of their work and market situation, and the ambiguity which surrounds their career line, has made it difficult for technicians (with the exception of draughtsmen) to discover an identity in an occupational interest-group. There has been little spontaneous effort to form either a single professional association or a single trade union catering for the whole group.[1] Without either the 'restrictive practices' of a union or the 'code of conduct' of a professional association, there has been little to prevent management from providing their own job specification to suit the task to hand.

The term 'technician' has been generally used to describe a 'catch-all' category of employee lying below managers and technologists at the upper limit and above skilled manual workers at the lower. Attempts at functional definition must ultimately be made with reference to one of these firmly established groups. The Committee on Manpower Resources for Science and Technology, in their report on the 1965 Triennial Scientific Manpower Survey, attempts to place the technician, both educationally and in terms of responsibility exercised in his work role, between managers or technologists on one hand and foremen or craftsmen on the other.

Technicians and other technical supporting staff occupy a position between that of the qualified scientist, engineer or technologist on the one hand and the skilled foreman or craftsman or operative on on the other. Their education and specialised skills enable them

---

[1] This is not to say that unions such as the Draughtsmen and Allied Technicians Association have not contributed to the early specification of the working conditions of their specialist group of members. Similarly some professional associations catering for both technologists and technicians have produced reports on training which come very close to specifying the types of job for which the training is intended: see for example the work of the joint Committee on Practical Training in the Electrical Engineering Industry set up by the Institution of Electrical Technician Engineers. These attempts at institutional constraints on labour supply are discussed in the next chapter.

to exercise technical judgment. By this is meant an understanding, by reference to general principles, of the reason or purpose of their work, rather than a reliance solely on established practices or accumulated skills.[1]

The needs of technicians

The implications of this statement of the position of the technician within the plant hierarchy has wider social implications. Much of what in wider society is regarded as appropriate to a person's life-style and behaviour stems from the position accorded to the employee within the work-place. Although many other forms of status and prestige may be associated with leisure or family roles, there seems little doubt that for many technicians their work status is extremely important in establishing their individual self-identity.[2] The Haslegrave Committee Report made extensive reference to the wider social status of technicians. It explicitly recognized that one function of nationally established educational qualifications would be 'to make some contribution to considerations of status for technicians'. In doing so it quoted from a speech made by the Minister of State for Education and Science at the 1966 Commonwealth Conference on the Education and Training of Technicians:

> The technician is neither a superior tradesman nor a depressed technologist. It is indeed essential that the technician be accorded a status of his own, that he feel himself to be a member of a body with an ethos of its own, a body of men—and women—who have wanted to become technicians rather than anything else, who have

---

[1] For example see the use to which socio-economic categorization is put in consumer research.

[2] This is a view which appears in the work of Alexander Wedderburn, *Work Orientations among Draughtsmen*, S.S.R.C. Conference in Social Stratification and Industrial Relations, Cambridge (unpublished); and K. Macdonald and W. A. T. Nichols, 'Employee Involvement: a Study of Drawing Offices', *Sociology*, Vol. 3, No. 2 (May 1969); as well as American Studies such as Louis H. Orzack, 'Work as a "Central Life Interest" of Professionals', *Social Problems*, Vol. 7, Fall 1959; and John W. Riegel, *Collective Bargaining as viewed by unorganized Engineers and Scientists* (Ann Arbor Institute of Labor and Industrial Relations, The University of Michigan-Wayne State University 1959). This view has been partially called into question in a paper by Professor Cotgrove in *Sociological Review*, July 1965, op. cit., in which he emphasizes the instrumental approach to work displayed among a polyglot sample of technicians at one plant. One conclusion stemming from the current study is that there are significant differences in orientation to work among various sub-groups of technicians.

been selected as having the right qualities for a technician, who have had the education and training appropriate to a technician, and who are proud to bear the title of technician.[1]

The Minister was himself quoting from one of the several conference papers reflecting concern with this issue. His statement followed a comment in the Report on Triennial Scientific Manpower Survey of 1965 on the urgency of providing technicians with a career line as well as a qualification which reflected their status.

There are few occupational groups in any society which enjoy the certainty of identity and status which the Minister was wishing for technicians. Generally these are long-established and well-recognized groups who have acquired some form of institutional control over entry to their work and training for it in the manner of the older professions and crafts. But there are equally well-established, but currently less prestigious and less well-defined groups such as managers and lesser skilled manual workers against whom technicians can compare their status—however crudely—in both financial and social terms. In the netherland in which technicians exist it is perfectly possible for them to seek an identity with either of these more established groups.

The small numbers in which technicians are usually employed, and their work situation, which mirrors an occupational location between that of manager or technologist and the shop-floor worker, makes it difficult for any collective occupational consciousness to emerge. In this situation it seems much more likely that the technician will resolve his personal uncertainties by identifying with one of these larger and better-recognized groups. In doing so his choice will depend upon his personal values and orientation to work. A recent Department of Employment and Productivity survey reported a group of technicians who were performing tasks considered by management to be those of technicians, but who refused staff status because they were able to earn more by means of a weekly payment-by-results scheme than on a monthly salary. On the other hand there is evidence that the acquisition of the title of 'technician' may bring about a substantial realignment of status goals for some former manual workers.[2]

[1] op. cit., p. 8.
[2] See Chapter 5.

The 'marginal man'

The technician accords with the now classical sociological description of the 'marginal man'.[1] In all societies, both locally and nationally, there is a structure of occupations in which there are clearly recognized bench-marks in the determination of a person's prestige and status. In most societies, and increasingly in rapidly changing industrial societies, there are wide areas of uncertainty in the society's evaluation of occupational roles, and as a consequence the individual actor's self-placement of his own position. It may be the sheer novelty of the occupational role that leads to uncertainty. Within more traditionalist societies newness will often lead to an incongruity between the values represented by the well-established bench-mark positions and those of the parvenus. This may lead to a delayed recognition of the modifications in the occupational structure. In the case of the technicians, for example, the real battle for acceptance was almost certainly that which was being fought by the technologists and would-be 'professional' engineers. The long-drawn-out struggle led by vocational educationalists against the defenders of sixth-form Classics and the 'all-round man' was one that had to be won before society in general, and higher management in particular, could concentrate its collective mind on the more lowly placed technicians.

The marginality of the technician's occupational role is often reflected in the inconsistencies in behaviour and attitudes displayed towards him both in the work situation and outside the plant. For instance, where the work role of the technician requires the exercise of authority over the production line, his essentially staff function, without the staff prestige accorded to the professional engineer, is an extremely difficult one to fulfil. The very ambiguity of his own position is such that these inconsistencies, particularly when shown by those whose respect he seeks, may be a constant source of irritation, frustration and even personal threat.[2]

This personal feeling of insecurity may not even require the display of what others might regard as disrespect. Marginal social groups of all kinds have been observed to provide a more than

---

[1] For a full description of the effects of social marginality see chaps. IV and V in R. K. Merton, *Social Theory and Social Structure*, (Free Press of Glencoe 1957).

[2] A fuller analysis of the technicians' reference group behaviour is contained in Chapter 5.

usually high proportion of deviants from prevailing social norms. Frustration may be expressed in any one of a number of ways but very often leads to aggression, retreatism, or innovative behaviour designed to remove or to avoid the social constraints to fulfilment of the individual's goals.

Yet even the most innovative, aggressive or retreatist behaviour may be accompanied by an immense desire to be accepted by one or all of the well-established groups to which the 'marginal man' aspires. Hence many aspects of his behaviour, particularly when interacting with this reference-group, may display a marked tendency to over-conform or to over-act the part perceived by him to be that of a true in-group member. (He becomes 'more British than the British'.) At one and the same time this ultra-conformist individual may be expressing his (currently) more truly characteristic behaviour in a different context. It is, for example, perfectly possible for a technician to identify with management in a way which would be unusual among manual workers, while expressing himself in collective action in which the leadership make demands of management with an aggression equally unusual among manual workers. At the moment, for many technicians a recognized path for promotion exists from the area of marginality to the well-defined and relatively prestigious position occupied by a technologist. The safeguarding of their present financial and even social status within the work-group through membership of a trade union, while they pursue their individual goals through their own efforts and possibly with the assistance of a professional association, may offer itself as a viable and acceptable dual strategy to such technicians.

Thus the concept of the 'marginal man' may provide some guide to the attitudes and behaviour of the technicians that are examined in this study.[1] It does not follow that all the

[1] ...The hypothesis being advanced is that the structural marginality of the technician's occupational status is such that in certain settings it leads to feelings of psychological marginality. Thus the ambiguity of his occupational role leads to behaviour on the part of others which is inconsistent with his own self image and to his expectations. The resulting state is known as one of 'cognitive dissonance' in which the party's failure to receive anticipated forms of response from 'important others' leads to uncertainty and a feeling of 'normlessness'. (Appropriate responses would for example include recognition for his long period of study devoted to passing higher examination grades than those used in his job.) The constraints of socially binding standards of behaviour may be eroded over a

behavioural characteristics of technicians can be analysed in these terms, nor does it mean that this rather crude analytical tool is not capable of much greater refinement in specific situational or personal cases. The very diversity of the work situations in which technicians are to be found demands more detailed analysis.

The way in which the technician views his personal situation will also be affected by many factors that lie outside or are little affected by the work situation; for example his age, family background, and leisure-time pursuits. It may also be true that in modern industrial societies 'acceptance' by those whom one regards as one's peers has to be expressed in a more tangible and universally recognizable form than a simple exchange of respect and responsibilities in the place of work. That is to say, status is seen as being expressed in largely financial terms, and the establishment of an occupational differential in the market is regarded as having precedence over more job-centred aims. This does not argue that Professors A. H. Maslow[1] and F. Herzberg[2] and other behavioural theorists who have placed their emphasis on the latter factors were totally wrong in their analysis. It merely re-states the old scientific axiom that 'all things are relative' and that the establishment of a 'sufficiency' or optimum of financial incentives (or 'hygiene' factors) may be a difficult exercise to carry out in a world of rapidly changing expectations. The belief that it was possible to establish an inherently fair structure of earnings and then set it aside to get on with the real job of 'motivating' technicians through changes in job-centred factors had brought considerable emotional traumas to some of the more 'progressive' personnel managers interviewed in the course of this study.

Neither should the argument necessarily have to be made in terms of a contrast between the 'traditional society', in which

---

considerable period during which the technician is 'over-responding' to the norms of the group to which he aspires (e.g. that of the technologist or middle management). Yet added stress derives from operating in his work task role under conditions which act to reinforce a feeling of 'injustice' and eventual alienation from those goals to which he once aspired.

[1] A. H. Maslow, 'A theory of human motivation', *Psychological Review*, Vol. 50, 1943, pp. 370–96.

[2] F. Herzberg *et al.*, *The Motivation to Work* (Wiley, New York 1959).

status or 'stand' was all important, and the new 'industrial' acquisitiveness. In the former type of society the extent and 'appropriateness' of one's wealth and income were known throughout the local (total) society. The real difference between the new and old forms is in the pace of social change, and therefore the ambiguity as to the 'appropriateness' of the income /status relationship rather than the importance attached to it. In the concern which technicians display towards their differential earnings their response is not so greatly different from the attempts by early craft associations to define their occupation in financial terms: *their* prototype was probably the earlier 'closed' guilds.[1]

Though this differential may be expressed solely in terms of differentials in immediate earnings, it seems likely that, unless there exist some strong external or ideological constraints, the technician will be highly motivated to maintain or secure some form of promotion ladder which will take him out of his present ambiguous status. Where such constraints as advancing age, lack of qualification or regressive earnings in higher positions detract from these ambitions they will clearly create problems which are analytically distinct from those of the younger, better qualified and less well-paid technicians.

In all of this the management response to the technicians' aspirations may be quite crucial. The recognition given to the technicians as an occupational grade may vary from establishment to establishment. In some cases technicians may feel no compulsion to bargain collectively with management, being satisfied with the recognition shown to them as a group through other means. In others the importance attached to promotion outlets and to personal relationships with management may be such as to deter such collective action; in these cases overt and deliberate intimidation on the part of management is hardly necessary. The sheer growth in numbers among technicians may have helped to reduce such barriers to a collective identity in a union. This does not imply that the status aspirations of the new technicians are any the less. It simply indicates that these and their individual career aspirations can be pursued in a

---

[1] See for example Sidney and Beatrice Webb, *Industrial Democracy* and *History of Trade Unionism*.

manner which is detached from their collective, and increasingly impersonal, contacts with management.[1]

The formal definition of an occupation and its effects

The new forms of training and education described in the last chapter may be regarded as part of this growing impersonalization, which has replaced the locally specific apprentice-training with a college-based selection procedure. In themselves they represent means of clarifying the technicians' position. However, they are at once goals and barriers; that is to say, goals for those who are young and/or able enough to attain the standards required in order to gain them, and barriers for those who are not. If, as the Training Boards and educational authorities expect, the industrial employers recognize such qualifications, then clearly the technicians' ability to identify with managers or technologists, or indeed with shop-floor workers, will be severely restricted. It may be a comment upon the restricted focus of most of the studies quoted in Chapter 1 that few of them considered the question of technician careers, as distinct from the 'current supply to the technician labour force'. Few of them considered the hopes and aspirations that motivated technicians in their occupational experience. Even the Minister of State hopefully saw the technician recruit as joining an occupation that was to be a terminal one: i.e., one in which his career began and ended as a technician. While this may be currently true of older technicians promoted from the shop-floor, it seems more likely that for many new young ambitious recruits it is a preparation and stepping-stone to a career extending to high occupational levels. In short, it is seen by many as a bridging stage, rather than a complete occupational career in itself.[2]

If this is so then on arithmetical grounds alone the chances

---

[1] For a discussion of the relative importance of these factors see David Lockwood, *The Black-Coated Worker* op. cit.; Kenneth Prandy, *Professional Employees: A Study of Scientists and Engineers* (Faber and Faber, London 1965); R. M. Blackburn, *Union Character and Social Class* (Batsford, London 1967); and George S. Bain, *The Growth of White Collar Unionism* (Oxford University Press, 1970). The analysis set out above differs in a number of rather important respects from that contained in these works. It is both extended and modified in Chapters 3 and 8 of this book.

[2] For a fuller typology and analysis of career mobility see J. H. Smith 'The Analysis of Labour Mobility', in B. C. Roberts and J. H. Smith (Eds.), *Manpower Policy and Employment Trends* (L.S.E. Bell, London 1966).

are that many will have to adjust their level of (realistic) expectations during the course of their career, since, although there appears to have been a growth in senior co-ordinating positions as well as, more lately, of positions among technologists, technicians still preponderate over technologists in industry. What is more, the increasing supply of graduates to fill technological posts (or to fill preparatory positions in the technician category) will have the effect of closing these promotion outlets to non-graduates. Many youngsters who would formerly have accepted the ill-marked route of the technician career-line are now clearly opting for the more direct academic path to a technologist qualification. Only if employers accept the social costs of providing a general managerial training for the intermediate 'technician engineer' grades will some of the plant-trained technicians gain a formal qualification suitable for higher positions. Competition may in any case grow fierce from other sources. As the general standard of administrative training for managerial and clerical workers grows higher, jobs which were formerly open to ex-technicians may close. (As an example of the present interchange in positions, in a number of plants surveyed in the present study, personnel managers were ex-planning engineers or production engineers who had been previously promoted from the ranks of technicians.)

The result of these pressures is the almost inevitable creation of a ceiling for the objective life changes of most technicians. That is not to say that the individual technician will accept this ceiling as being applicable to him personally. It has taken some three generations for technicians to begin to gain acceptance as an occupational group with their own interests and institutions. The constraints which this formal identity may place upon their individual career opportunities will probably not take them so long to discover.

Implications for the study

It has been suggested that many of the characteristics of technician behaviour stem from his ambiguous work and social situation. This in turn stems from his marginal status in the plant and wider social hierarchy. Normally, attempts at 'functional' specification refer to 'responsibility', 'understanding', and even 'supervision', without this being related to the status or authority

attached to the technician's role. Indeed, it is much more normal
for such definitions to speak of 'liaison' and 'communication',
and to place the occupant of the role firmly on an unspecified
base somewhere between management authority and shop-floor
power. It is apparent from the evidence to be set out in this
study that the average manager promotes technicians off the
shop-floor and regards their position within the firm as largely
instrumental to the short-term goals of the organization, without,
that is, attaching a great deal of formal status or concomitant
authority to the position.[1] As a result the tasks in which technicians were engaged and the titles under which such tasks were
organized were extremely varied and diverse. Not only was this
so, but new tasks arising out of changes in the technology, or
organization, were not slotted into any recognizable and systematic allocation of work-roles: they were simply added to the
existing heterogeneous mix. Since the technician was seen as
attaching some importance to his work-role in securing an
identity in a wider social context, and indeed in expressing his
own personality needs, the resulting frustration could in some
circumstances become a quite fundamental aspect.

Until recently little attempt has been made by Government,
industrialists or educationalists to impose a uniform framework
for recruitment and training to the occupation. In this respect
the work of these national agencies is increasingly being
influenced by hitherto weak professional bodies. It has, for
example, been observed that one of the recurring manifestations
of the emergence of a professional identity is a creation of a new
less-than-professional grade.[2] Thus, for example, in nursing, the
emergence of the State Registered Nurse coincided with the
creation of non-registered grades designed to take on the less
demanding work in terms of the acquired skills and ability of
the supervisor grade. In the creation of the 'technician engineer'
grade the Industrial Training Boards are doing no more than
following the advice offered by a number of professional bodies
catering for technicians and technologists, most notably the
Chartered Institute of Engineers.[3] In doing so they may succeed

---

[1] The empirical evidence produced for this study to support this inductive
generalization is reproduced in Chapter 5.

[2] cf. Ronald G. Corwin, 'The Professional Employee: A Study of Conflict in
Nursing Roles', *The American Journal of Sociology*, No. 66 (May 1961), pp. 604–
15.

[3] See Chapter 3.

in retaining a career line for the technician, leading into higher grades of work, which would otherwise go by default. As in the case of the nurses' bid for professional status, the pursuit of this end may imply the sloughing off of the unqualified 'dilutees' who, according to the previously quoted reports, make up the majority of those now regarded by management as technicians.

The initiative taken by the State and educationalists in the creation of an occupational identity is then, slowly but inexorably, being taken up by other interest-groups external to the technicians' work-place. These are for the most part organizations that were already in existence, which have extended or increased their interest in technicians' problems. On the one hand, organizations which have been seeking professional recognition for technologists have been made aware of a much larger audience among ambitious young technicians. On the other hand, trade unions catering for other occupations, or only for sections of the technician population, are canalizing the frustration and protest of this major group in society who have remained for so long 'marginal men'. In this field also the State has acted to support the embryonic trade unions. Clearly for these latter organizations the lesser qualified technicians, who are in the majority, provide a rich field of recruitment.

3 Growth and Development of Technicians' Institutions

Men and women join trade unions and professional associations for a variety of reasons. Each individual will be influenced in his decision by a personal set of values and psychological traits. Theories of trade-union growth have generally stressed the external influence of economic and political factors. The more recently developed understanding of the sociology of organizations suggests also that the internal structures, patterns of leadership and factors relating to administrative efficiency are important elements in the dynamics of union development.

Technicians, like other groups of employees, attempt to solve their individual economic and social problems by joining together to act collectively. Yet they may, as individuals, feel they have much to gain in a more positive way by joining an association in which membership is recognized as conferring status on those who belong. For whatever reason, technicians are the most unionized of the white-collar group of employees in industrial employment. This chapter will seek to examine the historical factors which have brought about this readiness to join trade unions, and the pattern of their collective behaviour.

Over the period 1964–68 the number of technicians increased by 38 per cent, and the number of technicians in the major unions catering for their specialized interests increased by 34 per cent; in the subsequent two years the number of technicians appears to have increased at the annual rate of approximately 10 per cent, but the growth in membership of technician unions rose dramatically by almost twice this figure. Yet it is difficult to measure the annual growth in any exact fashion because of the increasing number of organizations which describe themselves as 'associations' and have formal objectives which appear to differ from those of a trade union. Indeed all of the major technicians' unions, including the most notoriously militant, describe themselves as associations.

It is a title which in itself implies the desire of technicians to have their status as employees of a superior type to manual workers recognized. Belonging to an association, as distinct from belonging to a union, confers the status on technicians which is generally attributable in contemporary society to members of professions. In classifying technicians' organizations it is useful to begin by reference to the spectrum of collective organizations, which range from the trade union, which is entirely 'open' to any worker to join, however lowly his job status in terms of skill or wider social prestige, through those of the traditional skilled workers, in which limited entry is allowed to a craft-based organization, and finally to the traditional form of autonomous professional association.

The 'open union' depends upon the pure power of numbers to extract desired improvements in conditions of employment from employers, through the threat of a withdrawal of labour, and the disruption of the orderly process of work within the enterprise.[1] The union that is able to control entry to the job, in the manner of the traditional skilled workers' associations, is able to exert a greater pressure upon employers, and in times of a shortage of skilled workers may be able to obtain a considerable economic rent for the skills of its members through collective bargaining.

At the far end of the continuum is the professional association, which lays even greater stress on its ability to confer prestige upon its members than does the craft association. Usually this is achieved through the acceptance of a large measure of responsibility towards 'the community' in the maintenance of work standards. In order to fulfil this social function it will often seek a licence or charter to act as the sole provider of this service.

In order to gain this licence it has to achieve some sort of acceptance of its right to do so from the users of its service.

[1] H. A. Turner, *Trade Union Growth, Structure and Policy* (Allen and Unwin, London 1962). For an earlier development of this conceptual frame see Sidney and Beatrice Webb, Part II Chapter 2, 'The Method of Collective Bargaining', in *Industrial Democracy* (Longmans, London 1919).

C

If in fact consumers—say management—reject the qualifications of competence issued by the professional institution—no matter whether it has a licence or not—its monopoly is not likely to be maintained. In this case not only will its members lose their status in the eyes of others, but also their market rewards.[1]

Members of professional associations are usually very status-conscious. A high social status may be a substitute for the market power a union seeks to provide. Prestige within the place of work (or in any other situation) is usually accompanied by an appropriate authority over others or over one's own work. This in itself may be regarded as a satisfactory alternative to high pay. Professional people normally lay great emphasis on their need for autonomy at the place of work, and, if they fail to gain this autonomy, they may feel they have lost their status and seek to gain monetary compensation. A development of this kind may lead them to lose confidence in their professional association unless it can assert an ability to regain control over their work.

Yet in a highly industrialized society status is often measured by relative levels of salary, and a high level of responsibility within the work organization is seen as being accompanied by an 'appropriate' level of financial reward. This is increasingly the situation in Britain, where it is now quite normal to hear 'professional' people referring to relative reductions in income as reducing their status. Similarly, relative increases in the income of occupational groups whom they consider inferior in status to themselves may lead professional people to demand increases in remuneration. It is a common feature of the modern collective organization of professional people such as the British Medical Association or the Association of University Teachers to use status in an *instrumental* fashion. A claim to superior status which does not result in increased market power is rarely taken seriously by either the participants or the spectators, and strategies previously considered appropriate to that of a well-established occupational group, but which prove to be economically unrewarding, are gradually being abandoned in a rapidly changing social world.

The student of industrial relations is aware of this progression

[1] Many of the notions used in this chapter stem from E. C. Hughes, *Men and their Work* (Collier-Macmillan, Glencoe, London 1964), Chapter Six, 'License and Mandate'.

from concern with status to concern with cash in the histories of most craft unions. Hence the earliest associations of craftsmen attempted to maintain the status and life-style of their members through tight control over the job and the behaviour of the members. They set district rates of remuneration at which members could be contracted to individual employers and upon which they were reluctant to bargain. They attempted to control the supply of labour through a long period of training.[1] Unlike the successful professional associations, the training was not required by the technical needs of their trade to be of a depth sufficient to give its possessors control over their market position in the face of technological change. The history of modern technicians' unions is an excellent example of this weakness, while that of professional institutions provides us with an even clearer example of the erosion of 'control through consensus' within society at large and the consequent failure of modern associations to obtain and maintain a favourable market position for technicians through the method of autonomous job-regulation.[2]

Trade unions

Although technicians, with about 30 per cent of their numbers in unions in manufacturing industry, are the most unionized of white-collar employees, this figure hides a wide variation in the densities of organization of different types of technicians. In 1964 draughtsmen, who comprised almost half of the Department of Employment's 'technician' category, were 49 per cent organized. The 15 per cent of unionization in the 'other technicians' category was much closer to the overall white-collar average of 12 per cent.[3] There were also marked variation in the industrial distribution of union membership among technical staff, as

[1] For a summary of the historical evidence in the U.K. see Sidney and Beatrice Webb, *The History of Trade Unionism 1666–1920* (Longman's, London 1920), and for an account of the parallel development in the U.S.A. see N. W. Chamberlain and J. W. Kuhn, *Collective Bargaining*, second edition (McGraw-Hill, New York 1965).

[2] See *Industrial Democracy*, op. cit., for a conceptual development of this method as being situationally appropriate to 'The Doctrine of the Vested Interest' a doctrine developed on the basis of a high level of 'established expectations'.

[3] See G. S. Bain, *The Growth of White Collar Unionism*, (Oxford University Press, 1970), p. 34, Table 3.8.

well as across the range of technical occupations. Unions were most strongly organized among draughtsmen in vehicles (80 per cent membership), shipbuilding and marine (67 per cent in membership), engineering and electrical, and metal manufacture (50 per cent in membership). Other technicians were 37 per cent organized in timber and furniture, 28 per cent in shipbuilding and marine, and 27 per cent in metal manufacturing. However, in the intervening six years the proportion of 'other technicians' who are in unions has almost certainly risen, despite the continuing growth in overall technician numbers, i.e. in the potential pool of members. This is certainly true in the engineering and electrical industries for example, where the 17 per cent level recorded in 1964 has probably doubled. Indeed if one accepts union membership figures as a true record, then the 1 per cent membership in paper, printing and publishing, for example, or the 3 per cent in food, drink and tobacco must have increased several times over.

Up until recent years only two unions which specialized in servicing technical staff had met with any marked degree of success. Of these the Draughtsmen and Allied Technicians' Association (DATA)[1] is able to show densities of up to 80 per cent in industrial areas and in many plants in which it organizes. The other major union, the Association of Scientific Workers (A.Sc.W.), had a large proportion of its membership in the National Health Service and in Universities.

Since January 1968 the A.Sc.W. has been merged with the Association of Supervisory Staffs, Executives and Technicians, a union which until 1942 had limited its members to supervisory staff. At the end of 1969 the organization resulting from the merger with the A.Sc.W., renamed the Association of Scientific, Technical and Managerial Staffs, had a membership of 112,000, of which approximately half were technicians.[2]

With DATA membership standing at over 85,000, these two unions represented the largest organizers of technicians in the private sector at the end of 1969. There are, or have been, many more specialized groups of technicians, such as the Engineers'

---

[1] In early 1972 after the merger with the Amalgamated Union of Engineers and Foundryworkers (described on p. 110), the DATA again changed its title to that of the Technical and Supervisory Staffs' Section of the Amalgamated Union of Engineering Workers (T.A.S.S.).

[2] Source: *Report of the Chief Registrar of Friendly Societies for the year 1969*, Part 4 – 'Trade Unions' (H.M.S.O., London 1970).

Guild and the Association of Building Technicians. These associations have never met with any high degree of success even within the closed fields in which they have organized since the first world war. The growth of these and other organizations catering for technicians cannot compare with rates of growth achieved by ASSET, A.Sc.W. and DATA over the past sixteen years. Between the ten years 1955 and 1964 ASSET achieved a growth of 103 per cent and A.Sc.W. of 75 per cent; over the six years 1964 to 1970 the combined membership of A.S.T.M.S. rose by over 460 per cent to a claimed membership of over 200,000 at the end of 1970. DATA, which had achieved a growth rate of only 20 per cent over the ten-year period between 1955 and 1964, grew by more than 51 per cent over the period 1964 to 1970. Although more modest, the increase in membership of DATA can be almost wholly attributable to recruitment among technicians in manufacturing industry, indeed almost wholly within the engineering and electrical sectors. The more spectacular success of A.S.T.M.S. has stemmed as much from a broadening of its boundaries into clerical and administrative fields of employment as from the recruitment of technicians. We shall return to this point later.

There are at least a dozen other major unions involved in organizing technicians. In the public services there are the Post Office Engineers' Union (P.O.E.U.), the Institution of Professional Civil Servants (I.P.C.S.), the Society of Technical Civil Servants (S.T.C.S.), as well as the National and Local Government Officers' Association (NALGO). In the public sector there are occupationally closed unions organizing significant groups of technical staffs, such as the Electrical Power Engineers (E.P.E.), Association of Broadcasting Staffs (A.B.S.), Association of Cinematograph Television and Allied Technicians (A.C.T.T.) and, in the steel industry, the industry-based Iron and Steel Trades Confederation (I.S.T.C.).[1] The activities of both A.Sc.W. and DATA were curtailed within the public services by the 1927 Trades Disputes Act under which civil servants could no longer be represented by external bodies for the purposes of negotiating wages and conditions within the

---

[1] I.S.T.C. has, in the past, been referred to as the British Iron and Steel and Kindred Trades Association in fulfilling its negotiating role. This title was dropped shortly after the setting up of the British Steel Corporation in 1968.

Whitley machinery.[1] Their relationship with civil-service unions
has developed on a complementary basis, and since the Act was
repealed in 1946 there has been little attempt to compete with
the well-established public-service unions except in health and
education. The public-service unions enjoy a 'special relationship' with their employer, and membership densities of up to
90 per cent of eligible grades are normal within the national
Civil Service and local government.

The potential for recruitment in expanding occupational fields
is obviously very great, yet the unions organizing technical staffs
are quite old. DATA, for example, was founded in 1913, but
made little attempt to recruit outside the drawing office until
recent years, and unionization among other technicians remained
relatively slight in the private sector until recently. Some manual
unions have organized technicians because they had achieved
recognition rights that allowed them to offer a service to all
grades of workers within an industry (I.S.T.C. for example).
Other examples, such as the Amalgamated Engineering and
Foundry Workers' Union and the Electronics Electrical and
Plumbing Trades Union, have 'followed' members upon their
promotion into technician jobs. The response of these craft
but shop-floor-based unions has been remarkably different.

Until recently the A.E.F. paid little attention to its technician
members, for whom it does not possess recognition rights in the
engineering and shipbuilding industries. The E.T.U., on the
other hand, was more concerned when it changed its title in
1967 to Electronics Electrical and Plumbing Trades Union.
The first part of the title was designed to demonstrate its desire
to negotiate on behalf of technicians in the sectors it covered.
In 1968 the A.E.F. had something up to 700 supervisors and
technicians in membership; the E.E.P.T.U. had 2,200.

Clearly some unions have recruited technicians almost by
accident, simply because they were an available means of representation at a time when a group of technicians required such a
vehicle. The Union of Shop, Distributive and Allied Workers, for
example, offers a service to technicians in its Chemicals Division
which stems from its willingness to organize a group of discontented laboratory assistants in an oil refinery in 1940—after

---

[1] In 1969 it was announced that I.P.C.S. and S.T.C.S. were to merge since
the grades of civil servants which these organizations represented were likely to
be merged in line with the Fulton Report on the Structure of the Civil Service.

the Transport and General Workers' Union had refused to take them into membership. Such reluctance to organize technicians is not common today. At its 1969 Conference the T.G.W.U. decided to change the title of its Clerical and Administrative sections to the Association of Clerical, Technical and Supervisory Staffs.

In many instances these alterations have been made in response to technological changes in the industries organized by these unions; technological changes which have had a profound impact on the occupational structure of the industry and also, in many cases, on the career expectations of existing members. But the union response is generally somewhat delayed, and the obstacles to recruiting technical staff often lie in the attitude of members and leaders of manual unions.

The change of organizational style required is nevertheless sometimes quite difficult for manual unions to accomplish. Status and earnings differences between technicians and shop-floor workers may play a large part in the initial desire of technicians to join a representative organization,[1] and the belief that manual unions cannot be sufficiently representative of their 'special interests' is often a reason for not joining a trade union. For that reason the white-collar sections of more general unions have the initial difficulty of establishing a new and quite distinctive identity and style. Thus the Society of Graphic and Allied Trades' recruiting leaflet features the photographs of managing directors who were once members of the union, but ends with the discreet message that 'applications will be treated in strict confidence'.

While it is clear that many unions are now beginning to realize that unless they recruit technical staffs they will decline, as changing technology will erode their traditional bases of membership, technicians themselves view union membership with an equal degree of uncertainty. For many of them who are promoted from skilled manual occupations, continuation of union membership may be a natural proceeding; for others, however, promotion may give rise to a desire to seek confirmation of the attainment of a higher status by achieving membership of a different body, possibly one that clearly stands as a 'technicians' association'. A change of organization may, of course, be made necessary by the existence of procedural agreements concerning repre-

[1] See Chap. 8 for a fuller extension of this hypothesis.

sentation that prevent a manual union from offering a representative service to white-collar members. Technicians who achieve
their occupational status directly through technicians' training
schemes may well be reluctant to join a predominantly manual
workers' union. Their preference is likely to be for an exclusively
technicians' union, staff or professional association, or even no
membership of any group save that which confirms the
managerial or technologist status to which they individually
aspire. This latter category is clearly very large; its extent
depends mainly on the ability of unions and associations to
attract technicians into membership by an appropriately appealing range of activities.

Staff associations

Unlike the trade union or professional association, the staff
association is usually limited in its origins to a single enterprise,
and concerned with internal rather than external matters. Its
primary purpose is normally stated to be that of improving
communications between management and employees in
company decision-taking. Although the existence of a staff
association suggests the need for an institution through which
collective views can be obtained, and the possibility of differences
of interests between staff and management, it also suggests a high
degree of common interests and a desire to establish a relationship that is based on a unitary purpose rather than an inevitable
conflict. In fostering and using staff associations to achieve
these goals, management may be attempting to frustrate the
activities of external organizations seeking to fulfil a similar
representative function.

The history of staff associations seems to suggest that there
has been a strong relationship between the first appearance of
an external protective organization and the setting up of an
internal consultative body within the enterprise. This management strategy is designed to appeal to the desire of technical
staffs both to have a collective voice and to identify strongly
with management.[1] One major influence upon the organization
of staff associations is of course the size of the company. With
the growing size and concentration of firms employing tech-

[1] cf. Lloyd Reynolds, *Labor Economics and Labor Relations* (Prentice-Hall,
Englewood Cliffs, N.J., 1956).

nicians, the conditions likely to promote interest in such associations have become more widespread. However, in the 1890s, at the time of the formation of the first independent 'unions' in the shape of Foremen's and Technicians' Friendly Societies, the size of most engineering companies led employers to seek a broader collective response in the formation of a national body known as the Foremen and Staff Mutual Benefit Society, sponsored by the engineering employers in 1899.

In few respects did this Society meet the need for improved communications, and, when required, for effective representation. At the time of this study it existed to provide pensions, life assurance and sickness benefits to foremen and similar grades of staff in the engineering and shipbuilding industries. Society membership was made up of employers who pay at least half of the total contributions in respect of the second category of membership, their foreman employees. Its constitution contained a clause specifically prohibiting its employee members from belonging to a trade union. If an ordinary member (employee) wished to join a trade union after being received into the Society, he had to forfeit his claim for the employer's contribution made on his behalf, and stood to lose over half of the ultimate value of total contributions.

After receiving evidence on this aspect of the Society's function from all trade unions with an interest in middle grade staff, the Royal Commission on Trade Unions and Employers' Associations gave it as their view that 'it is quite foreign to the purpose of a Friendly Society that it should prescribe in its rule that no-one can be a member and draw benefit if he is a trade unionist'. The Commission therefore recommended that no Friendly Society should have such a rule.[1] Its recommendation was carried into effect by a Bill promoted in the House of Commons on a private petition of the Association of Scientific, Technical and Managerial Staffs in December 1968. In the event, the anti-union provisions of the F.S.M.B.S. were erased before the Bill went to its third hearing.

As an anti-union organization, the Foremen's and Supervisors' Mutual Benefit had a wide measure of success: a success denoted by the attention paid to it by A.S.T.M.S. Its membership in 1967 stood at over 61,000 ordinary members; for the most

---

[1] *Royal Commission on Trade Unions and Employers' Associations, 1965–68: Report* (Cmnd. 3623, H.M.S.O., London 1968) paragraph 252.

part these were foremen, and only a small number were supervisory technicians. The fact that the Engineering and Shipbuilding Employers' Federations advised member firms to join the Society and to encourage eligible staff employees to do likewise created widespread bitterness among trade unions attempting to recruit or retain membership in this area of employment. Their cynicism was reinforced by a reported decline in recruitment of new members to the F.S.M.B.S. after 1969.

It is possible that by the sponsoring of the Foremen's and Supervisors' Mutual Benefit Society employers may have contributed to the ambivalence or outright antipathy towards company-oriented schemes of consultation and company-encouraged staff associations reported in a later chapter. For the present it is perhaps worth recording that the existence of company staff associations for technicians alongside trade unions representing manual workers within the same plant seemed to contribute to a growing predisposition towards joining technicians' unions on the part of these involved in these arrangements. It was clear that these employees developed an ability to compare the effectiveness of their form of group representation in what was essentially a form of collective consultation with that provided by manual-worker unions in their collective *bargaining* relationship with management. This made it crucial to management that the comparison should be a favourable one; needless to say it rarely was.

It has subsequently been demonstrated that this process of unionization through staff associations is possible in a much wider context. A number of company staff associations within banking and insurance have seceded from their attachment to management. Some, like the Prudential Insurance Staff Association and Royal Insurance Group Staff Guild, have transferred their engagements directly to the A.S.T.M.S., while others, such as the Eagle Star Staff Association, have registered as trade unions. It would appear that once an identity of purpose has been established within such a white-collar interest-group a need to demonstrate an independence from its progenitors in senior management may develop.[1]

---

[1] The establishment of a new office of Registrar of Trade Unions and Employers Associations has brought changes in the law relating to registration of worker organizations. This has brought about an increase in the number of staff associations declaring themselves as trade unions.

Professional associations

The desire to be accepted by management according to existing standards of what is proper conduct in the determination of their conditions of work is obviously very strong among technicians. Professionalization may be seen as a desire on the part of members of an occupation to obtain status through attempting to organize an association designed to enlist or to retain high-prestige rewards from society at large or from some significant part of it. This enables its members to achieve or retain an autonomy within the place of work and to defend or secure an advantageous labour-market position.

The term 'chartered engineer' has come into currency only recently with the acquisition by the newly formed Council of Engineering Institutions of a Royal Charter. The establishment of the Council marks the high point in the professionalization of engineering in this country; as a profession, engineering has not had quite the same social standing as the law, medicine, or even architecture. Engineering has been characterized by the proliferation of specialist institutions which have sought to give professional recognition to a particular branch of engineering, and it was no mean feat to succeed in getting fourteen of the 'senior' bodies to agree to common standards of entry into the ranks of the professional engineers. Some of the oldest of these bodies, such as the Institute of Mechanical Engineers, date back to the first half of the nineteenth century. But in each decade over the past century there has been an increased number of specialist institutions formed in engineering—more than in any other definable profession. At present the fourteen members of the Council of Engineering Institutions are:

The Royal Aeronautical Society
The Institution of Chemical Engineers
The Institution of Civil Engineers
The Institution of Electrical Engineers
The Institution of Electronic and Radio Engineers
The Institution of Gas Engineers
The Institute of Marine Engineers
The Institution of Mechanical Engineers
The Institution of Mining Engineers
The Institution of Mining and Metallurgy
The Institution of Municipal Engineers

The Royal Institution of Naval Architects
The Institution of Production Engineers
The Institution of Structural Engineers

The hope is that the Council will act as the spokesman for the engineering profession as a whole, and will enable the public to have a clearer conception of the professional engineer than was the case when he was spoken for by a large number of separate organizations.

In 1969 the total membership of the C.E.I.'s constituent institutions stood at about 275,600, of whom about 116,000 were students and graduates (i.e. not full members).[1] Not all those who would be eligible for professional status apply for it, however, so that this figure represents a small fraction of the total population of potential professional engineers. Some estimates place the percentage of professional engineers who are not members of a 'senior' professional body as high as 60 per cent, though no reliable figures are likely to be available. In terms of the number of *institutions* that are likely to be eligible for C.E.I. membership, however, the position is more certain:

> The present constituent membership of the Council represents a very large part of the profession and without duplication of representation there are very few additional bodies that can be expected to qualify.[2]

In setting criteria for membership which guarantee exclusiveness, the C.E.I. is attempting to consolidate the status of 'chartered engineer'.

In more explicit terms, the C.E.I., in common with all bodies which seek to turn occupations into professions, has:

(a) restricted entry to those who satisfy the standards set up by the association;

(b) attempted to gain community recognition as the spokesman for a particular speciality and as the guardian of certain standards of entry and competence.

The achievement of a Charter in July 1967 set considerable problems for aspiring technicians having student membership of one of the fourteen bodies. Generally the standard of entry had rested at about Higher National Certificate level for most existing members of the constituent bodies. Those people who had

---

[1] From an interview with a national officer of the C.E.I., July 1969.
[2] Statement by Council of Engineering Institutions, November 1966.

gained membership in this way were allowed to gain the title implied by the Charter, and indeed, under Rule 33 of the Bylaws, 'corporate members of constituent bodies may be registered up to January 1974 on the basis of no less standard than was required to secure corporate membership in August 1965'. This condition applied particularly to those in the Society of Electronic and Radio Technicians and the Institution of Electronic and Radio Engineers in which a large number of members were technicians. After 1973 these people will become a dwindling proportion of those entitled to describe themselves as 'chartered engineers'. After that date corporate members will be recruited through a degree-level examination, and should be employed in a technologist capacity.

Other smaller organizations claimed professional status for technicians as distinct from technologists. These included the Architectural and Allied Technicians (created by the Royal Institute of British Architects), the Institute of British Foundry-men, the Society of Instrument Technology, the Institution of Plant Engineers, the Society of Engineers, the Institute of British Engineers, the Institute of Engineering Inspection, the Institute of Engineering Designers, and the Institute of Work Study Engineers, to mention the largest of over 100 small institutions claiming to represent professional technicians. Some of these were actually formed by older bodies, such as the Royal Institute of British Architects, to represent junior or older less-qualified people in their profession. This practice is not unique to engineering. It has been observed that the older professions and even in craft organizations high-status members may very often attempt to preserve their differential position by taking less-qualified people 'working at the trade' into a form of second- and even third-class membership of their organization, or starting a satellite body designed to prevent the 'technician' group from forming a separate, and possibly threatening, identity of its own.

In December 1967 the C.E.I. took the initiative in bringing together thirty of the smaller technicians' societies to discuss a confederation similar to that of the C.E.I. in respect to technicians. Following a meeting of these and eleven other bodies in May of that year, a scheme for federation, based round a register of 'professional technicians', was decided upon. It was deter-

mined that there should be two grades, the more senior based
on the H.N.C. qualification or its equivalent (Board of Trade
or City and Guilds examination), and that this grade should be
entitled 'Technician Engineer' to distinguish it from the lower
one. This second 'Technician' grade would be based on the
O.N.C. qualification and would be regarded as half-way house
to Technician Engineer.

The Engineering Industrial Training Board gave its approval
to these titles shortly afterwards. The major technicians' unions
expressed their discontent at the implied second-class status
given to many of their members or potential members. None of
them took any effective action, however, and the phasing out of
non-qualified members of the forty-one Technicians' institutions
began in 1968 over a similar seven-year time period to that of
the C.E.I.

The stage appeared set for a clear division of interest between
those with Technician Engineer qualifications who might adopt
the unilateral bargaining stance of the professional body and
those without such universally acceptable skills who could join
trade unions and bargain on sheer 'strength of numbers'. However, within twelve months of the granting of its Royal Charter
in 1967, an article appeared in the C.E.I. journal *Engineering
News* which pointed out that, whereas the professional engineer
had responsibilities to the public, his profession, his employer
and himself, in a situation of increasing insecurity the responsibility to himself was becoming increasingly important; 'the
rationalization of the pattern of industry has generated a feeling of insecurity and the need for collective protection ... and it
does seem that collective negotiations for more pay are no longer
beneath the dignity of very many professional engineers.'
Renewed activity was reported in the long-dormant Engineers'
Guild (membership 6,000), and within the space of a few months
individual members of the C.E.I. who were also Guild members
announced the formation of the United Kingdom Association
of Professional Engineers (U.K.A.P.E.). The Association was to
take over the negotiating functions for which the Guild had been
founded in 1930, and to recruit from membership of the C.E.I.

In June 1969 a second body, the Association of Supervisory
and Executive Engineers, was formed, which constituted a direct
rival to the United Kingdom Association of Professional

Engineers. 'However it will cast its net wider recruiting any chartered and non-chartered engineer holding managerial, executive and sometimes supervisory posts ... it could potentially have 750,000 members.'[1]

In the following month the Council of Engineering Institutions announced that it was to review the needs of members, and would canvass opinion on the reconstitution of the basis of its qualifying and learned-society aspects in favour of protective functions through more direct means. The move was said to be induced by a gathering insurrection among young engineers.[2]

By August 1970, when membership of U.K.A.P.E. had risen to only 9,000, a *modus vivendi* had been reached with the C.E.I. based upon the three-tier model of organization used by doctors in the Health Service. The C.E.I. is to perform the 'learned-society' functions of education and the setting of professional standards. A new body, the Professional Engineers Association Ltd (PEAL), offers legal advice, and mediation and arbitration between members of the profession and their employers. It also offers an appointments bureau of situations which 'conform to [PEAL's] policy or model conditions'. Thirdly, U.K.A.P.E. has now been registered as a trade union. It entered its first recognition dispute at C.A. Parsons' Newcastle turbine plant in 1970.[3] In this dispute and in the promulgation of its general strategy, it has to contend with the already well-established technicians' unions, who have recruited up into the ranks of technologists, as well as with the reluctance of the Engineering Employers' Federation to commit themselves to separate negotiations with yet another trade union, especially one that includes some very senior staff in its membership.

Perhaps the greatest problem confronting the new tripartite professional organization is that whereas members of, for instance, the medical profession, however scattered, are ulti-

[1] *Financial Times*, 12 June 1969.

[2] Statement of C.E.I. as reported in *The Times*, 14 July 1967.

[3] John Fryer, 'The "professionals only" union makes a bid for power', *Sunday Times*, 16 August 1970. This jurisdictional dispute which involved UKAPE in a direct confrontation with DATA has subsequently gained added significance by becoming the subject of the first reference to the National Industrial Relations Court under Section 45 of the Industrial Relations Act 1971. The company C. A. Parsons Ltd., sought the Court's guidance on the organization of bargaining units and which worker organizations, if any, should be recognised as sole bargaining agents for those units.

mately dependent on one major source of employment, the National Health Service, engineering employers are multi-various in their pursuits and in their market positions, even more so in fact than are the highly heterogeneous collection of pursuits covered by the 'engineering profession'.

In January 1970 the Council of Science and Technology announced[1] the formation of a Guild of Professional Scientists and Technologists to cover the 60,000 members of professional institutes in biology, chemistry, mathematics, metallurgy and physics.

In the following September the Royal Institute of Chemists announced that it too had determined upon a course of setting minimum rates and of advising graduates against accepting jobs in establishments where wages and conditions were below those approved by the Institute.[2] The fact that a large proportion of its 27,500 members are employed within two or three large combines encourages the Institute in its belief that it can handle salary negotiations on their behalf. There is, of course, a legal contradiction involved in the use of collective bargaining by a professional institution, since the monopoly powers awarded by Royal Charter cannot be maintained by a body engaged in anything other than the pursuit of the public well-being. The structural modifications introduced by the C.E.I., based on the model provided by the British Medical Association, enables such bodies to adapt their organizations to a two-pronged strategy in which the same officer often adopts different legal identities. But this does not allow for the possible psycho-logical dissonance created within the minds of members— particularly some older members—to whom the Charter has more than a simple market connotation.

However, the C.E.I. claim to sole professional status has already been challenged on its own grounds, and by the technicians themselves. After two years of debate the Council applied to the Privy Council to enable it to establish separate *professional* registers for chartered, non-chartered, and technician engineers. Since this attempt at rationalization had excluded a large proportion of the hundred or more remaining existing technician bodies, it seemed hardly surprising that in the same month (July 1970) twenty-three non-chartered institutions should announce

[1] *The Financial Times*, 16 January 1970.
[2] *The Times*, 28 September 1970 and 27 October 1970.

the establishment of a rival register by a body to be known as the Standing Conference for Technician Engineers and Technicians.[1] Some of the institutions have already met with success in the establishment of courses of study approved by the Department of Education and Science.[2]

The 'instrumentality' of status

The original attitudes of the founders of the Council of Engineering Institutions were ambivalent in that they ignored the importance of the labour-market connotations of the qualifications over which they had gained a monopoly by the granting of a Charter. The engineering institutions had followed an evolutionary path common to many of the more traditional professional institutions. They had first set up as learned societies and adopted this pattern of association within the new industries, such as chemicals, as they developed. The interests covered by most of the institutions were too narrow to form a viable basis for a 'closed occupation', and it was necessary to establish a federation prior to applying for a licensed monopoly over occupational qualifications. When this was acquired it became possible for the Council to replace such national qualifications as H.N.C. and university degrees with its own associate membership examinations.

However, such evidence as exists suggests that an independent professional qualification is no longer seen, especially by the younger engineers, to provide sufficient protection within a large modern bureaucratic organization. Professional engineers have come increasingly to believe that their qualifications are more than just a means to 'do good' within society or to maintain certain self-set standards of workmanship. They expect professional associations to be active in securing the higher financial rewards which have traditionally been regarded as appropriate to their social status. Over the latter part of the 1960s professional engineers fell behind in the upsurge of the earnings spiral; their salary levels even dropped behind those of pure scientists.

[1] *The Times*, 23 June 1970 and 8 July 1970.
[2] *The Times*, 7 November 1969.

In general the earnings of non-graduates increased more rapidly than those of graduates.[1]

In these circumstances it became increasingly difficult to convince younger graduate engineers that the professional strategy alone was as efficacious as that of collective bargaining. The apparent devaluation of their qualifications has come at a time when they are increasingly dependent on the large corporations, which are free to expand or to 'rationalize' their staff without reference to the professional institutions. An apparently successful negotiating organization like the Association of Scientific, Technical and Managerial Staffs offers an attractive alternative to those who are dissatisfied with the traditional role of the professional bodies.

For the technicians themselves the 1960s was a period in which their relative status was brought home to them, first by the total 'closure' of the chartered institutions', and later by the grudging acceptance of a small élite group of 'technician engineers' into 'third-class' professional membership. The ambivalence displayed by the unions towards this process could be interpreted as being entirely calculative, since they might stand to gain by this policy. However, the history of these unions (or staff associations) displays such a close parallel to that of the evolution of the modern professional bodies that it is possible that the leaders of white-collar unions see little potential rivalry in the present form of professional body.

From unilateral regulation to collective bargaining

Those unions which specialize in the recruitment and servicing of technician employees have usually begun their life by attempting to define and to isolate the occupational identity, the status and the market value of the groups whose interests they seek to represent. In order to do so there has had to be some embryonic consciousness of a separate occupational identity within the work and/or the market situation of potential members. This process is to be discussed in Chapters 5 and 8. But whatever the level and quality of expectations shared by

[1] See 'The Survey of Professional Engineers, 1968', *The Survey of Professional Scientists, 1968*, Ministry of Technology (H.M.S.O., London 1970). For the Guilds' Response see *Professional Engineering Responsibility Levels* and *Recommended Salary Levels for Professional Engineers* (Professional Engineering Publications, Liverpool).

work-groups of a similar status, the availability of a representative organization which reflects their common feelings and aspirations and which is prepared to give such goals a high degree of significance may attract support, although in the undertaking of collective action this may be in conflict with the personal ambitions of individual members. However, it may be possible for the individual technician to retain a personal allegiance to the work system of authority from which he derives his career opportunities, while seeking to ensure the collective status and security of the group he belongs to through membership of an external organization. If either career opportunities become diminished or the standards of expected professional conduct change, experience has shown that it is likely that the professional form of representation will give way to a more 'unionate' method of bargaining. It is worth while outlining briefly the historical evolution of the major technician unions in order to illustrate the movement from the closed (or vested-interest) membership policy of the early associations to the more open (collective-bargaining) position that they hold today.

The Draughtsmen and Allied Technicians' Association

Up to 1961 the title of this union was the Association of Engineering and Shipbuilding Draughtsmen. It is a title which more accurately reflects the membership of the organization over most of its history than does its present one. Its evolution as an organization has been structured by the market, technological, and political factors that have influenced the growth of industrial relations in the British engineering and shipbuilding industries during this century. Nevertheless it has specialized in representing an occupational group which stretches across several industries. Draughtsmen, more than any other category of technicians, have developed a craft or professional identity which distinguishes them from other categories of employees in any one industry.

(1) *The structure of DATA*

'The basic unit of organization in DATA is the office in which members are employed.'[1] The occupational identification of

---

[1] *Royal Commission on Trade Unions and Employers' Associations*, Minutes of Evidence No. 36, DATA (H.M.S.O., London 1966).

draughtsmen has a significance which goes beyond the sharing of common skills and a common market position. Few skilled craft or professional employees work in such close and permanent proximity to their fellows and share the same experiences and responses to the rest of the industrial system. In all that has been written on DATA, great stress has been placed upon the importance of the work-place group as the basic organizing unit, connected, through its locally elected 'Corresponding Member', to a formal co-ordinating centre in the national office.

The office work-group is not usually large enough to constitute a branch. Branches are related to the work location of the members, but attendance at monthly branch meetings may be a function delegated to office committee members or other activists; attendance of ordinary members is sparse.[1] Each branch is entitled to send a delegate to one of fifteen Divisional Councils. As in other craft unions, the Divisional Councils have a special responsibility for maintaining the 'district rate' in all negotiations with employers within their area. It is therefore the practice for the full-time divisional organizers to report to each of the Divisional Council's monthly meetings on negotiations conducted during the previous month.

In common with other unions, DATA regards its annual national conference of branch delegates, known as the Representative Council, as the supreme policy-making body; in practice the main work of the Council consists of the consideration of the Executive Committee Report, and motions and amendments from the branches. The Executive Committee is responsible for the day-to-day running of the Association. Since most of its members are lay members directly elected by branch ballot, its status *vis-à-vis* the Council is a relatively strong one. On the other hand the Council reserves to itself the power to appoint the General Secretary and five other national officers, together with the fifteen Divisional Organizers. The national President, Vice President and Treasurer are elected for periods of one to three years, but the General Secretary and all other headquarters and divisional staff are permanent appointments.

Thus the formal division of power within DATA has much in common with the fine balance of the nineteenth-century craft association, a fact which has had considerable bearing on its later

[1] In its Royal Commission evidence, DATA estimates normal monthly attendance at an average branch meeting to be 15, or just over 4 per cent.

development.[1] Yet in spite of the struggles for national power which have punctuated the union's history, the relationship between the office work-groups and head-office staff has remained crucial to the smooth running of the organization. These groups have formed a network throughout the engineering and ship-building industries. Headquarters has been the hub of a web of inter-office communications, and the co-ordinating centre for 'pattern bargaining' throughout the country along lines which can only be laid down by Head Office.

(2) Origins

The Association of Engineering and Shipbuilding Draughtsmen was formed in 1913. Most of the earliest aims of the A.E.S.D. did not depart radically from the 'friendly-society' objectives of earlier local associations of engineering foremen and draughts-men. The leading speaker at the inaugural meeting expressed prevailing opinion in an article written shortly afterwards. 'Amongst the majority of draughtsmen there has long existed a deep-rooted objection to rampant trade unionism, its despotism, its iron-bound regulations and its strife creating propensity ... any combination amongst ... must not interfere with the liberty of the subject.'[2]

The employers on the Clydeside had come to a 'gentlemen's understanding' not to compete with each other for the services of draughtsmen, and it was their action which had persuaded draughtsmen of the need to establish a nationally based trade union. This was a period of rising prosperity in which draughts-men could gain rapid wage increases by moving between firms. It was also a period of technological and organizational change, in which technical drawings were becoming less artistic and more standardized. A draughtsman's capabilities could there-fore be advanced primarily through his knowledge of a wide range of varying technological processes. Mobility was an increasing characteristic of the ambitious draughtsman; blocked

[1] DATA's leaders have, from the time of its origins, been strongly influenced by the workings of the Amalgamated Society of Engineers, the forerunners of the Amalgamated Union of Engineering Workers. In its formative years the members of A.S.E. were pushing through fundamental constitutional reforms designed to reduce the power of their Executive Committee.
[2] J. E. Mortimer, A History of the Association of Engineering and Shipbuilding Draughtsmen (A.E.S.D., 1960).

mobility served to bring home his position in the labour market.

From the inaugural meeting onwards, the Association attracted many senior draughtsmen to its ranks. These senior members have always tended to be pragmatic in their approach to unionism, and while favouring collective bargaining have laid stress on the need to maintain 'professional standards'.

(3) *Recognition and growth*

The first major negotiations entered into by the A.E.S.D. took the association no further than the actions of most professional bodies. Yet their effectiveness probably contributed to the dramatic influx of membership between 1914 and 1917. During these first three years of the first world war, membership increased from 300 pioneering members in and around Glasgow, to 10,000 members scattered in drawing-office combines throughout the country. In 1917 Headquarters were moved from Glasgow to London, partly as a result of geographical expansion, but also because the first major body to recognize the A.E.S.D. as a legitimate representative of draughtsmen had been the government departments in Whitehall. For the period of the war, responsibility for awarding national wage increases rested with the Ministry of Munitions, and the success of the Association in gaining a 12½ per cent increase after skilled manual workers had received a similar award gave it considerable stature in the eyes of draughtsmen. Its role as advocate in defending the mobility of members against the imposition of certificates preventing them from changing their employment was also important in bringing about a book membership of 14,384 by 1921.

The Association was recognized for the purposes of representing its members in grievance procedure by the Engineering and Allied Employers' National Federation. In 1924, after two long-drawn-out local strikes to which members gave their total support, the employers' leader, Sir Alan Smith, explained that the Federation's constituent members had wished to maintain close relationship with their own employees, but since A.E.S.D. had chosen 'another course' this was no longer possible. Having refused to recognize the A.E.S.D. between 1918 and 1920, during its more pacific formative years, the Federation were now anxious to discuss an arrangement for the settlement of grievances.

This was signed at a moment when the immediate rush of

Association recruitment was over—when in fact membership had been in decline for two years. The Employers' Federation, on the other hand, had just won a significant victory in the lock-out of A.E.U. members enforced in 1922. Though recognition was confined to non-management staff (and it still remains so), and excluded apprentices, it stands as proof of the success of the militant plant bargaining which had preceded it. This tactic of gaining a concession at a number of key plants and then propagating its gain through a national agreement was not to become the hallmark of DATA strategy again until the full-employment conditions after the second world war.

The A.E.S.D. came under increasing labour-market and political pressures during the late nineteen-twenties and early 'thirties. Recruitment to the union came to depend on the attraction of high unemployment benefits, maintained as part of the early 'craft' tradition. The shipbuilding industry was in an extremely exposed position in the inter-war years owing to the decline in the overseas markets. Despite their early initiative, the Association's members in shipbuilding did not adopt an aggressive position. Only two approaches were made to secure recognition at the national level by the Association. A third refusal from the Shipbuilding Employers' Federation in 1940 led to a request for the intervention of the Minister of Labour, Ernest Bevin. For the second time in its history the A.E.S.D. gained 'recognition' through the intervention of the Government.[1]

(4) Craft association or trade union?

The rapid expansion in membership which took place during the first world war was repeated on a greatly enlarged scale during the second. Membership of A.E.S.D. expanded from 19,310 in 1939 to 36,661 in 1945. After a long period of depression the return to a high rate of growth had taken place in 1936. The recovery of the engineering industry and the demands of rearmament brought with them a huge expansion in demand for technicians, and in particular draughtsmen. Much of the new growth took place in the South-East and Midlands. At the 1937

---

[1] For an interesting historical analysis of the significance of employer recognition see G. S. Bain, *The Growth of White Collar Unionism*, op. cit., or Research Paper 6, *Trade Union Growth and Recognition*, Royal Commission on Trade Unions and Employers' Associations (H.M.S.O., London 1967).

Conference a Scottish delegate proposed a committee to investigate the introduction of a control scheme to protect members
from the 'perils of dilution'. The committee, which largely consisted of delegates from London and Birmingham, presented a
report in 1938. It recommended that, since the DATA minimum
rate was being obtained by the 'dilutees' as well as by journeymen, no action should be taken.

This incident, which preceded many similar debates during the
second world war, serves to illustrate the dichotomy which had
existed between the positions of two groups of memberships
within the Association since the first great influx of 'unqualified'
dilutees during the first world war. It has given rise to an internal
dialogue on bargaining policy, paralleled by an ideological
division, which has underlain the argument chosen to legitimate
and justify the chosen labour-market strategy.

Basically the choice has been between the two positions
described at the beginning of this chapter, i.e. between the 'open'
trade union recruiting 'all working at the trade within a work-
place'[1] and seeking to impose a common minimum wage by
means of collective bargaining; and the 'closed' craft or professional association, whose concern has been to restrict entry to
the *occupation* through maintenance of work standards and so to
retain a market value for skills embodied in a required
apprenticeship. For these latter organizations, collective bargaining was normally to be regarded as establishing a grievance
procedure, a 'safety net' against failure by an employer to accept
the self-evaluation of the draughtsmen in his employ. For example, as late as the 1939 Representative Council, a proposal
that the Association should seek agreement with the Engineering Employers' Federation on minimum rates of pay was opposed
by the Executive Committee on the grounds that it was
'dangerous to the standing of the Association'.

Some of the interviewees in the survey looked back at the
days before the 1940s 'when the old A.E.S.D. was a professional
body'. Letters to the union journal often express a similar view-
point.[2] This, in fact, it never was. Among the aims of its founders
was 'to ensure that all who entered the profession were properly
trained [because] development of specialization . . . might tend to
lower the qualifications of draughtsmen unless proper precautions

---

[1] S. and B. Webb, *Industrial Democracy.*
[2] See DATA Journals during 1960s concerning debate on amalgamation.

were taken.'[1] At the 1917 Conference, which decided to seek registration under the Trade Union Act 1871, the alternative of seeking a Royal Charter as a professional association was strongly canvassed by a small group of delegates, but without any real chance of success.

The Association has, however, never lost its interest in control over training and work standards. In an Executive Report on Apprenticeship presented to the 1924 Conference it has suggested that the Association should conduct its own examinations and establish these as being the appropriate ones for journeymen. In 1935 the proposal was modified to accord more with reality by providing for 'joint regulation' of an apprenticeship scheme. In fact the Association achieved no more than normal negotiating rights for apprentices and, thirty years later, it acquired representation on the appropriate Industrial Training Boards on the same basis as that of other interested trade unions.

'Closure' of the occupation in a 'professional' sense was, therefore, never accomplished. Its various classes of membership are now based on experience 'in the trade'. But the Association has retained one important link with the 'learned-society' aspects of its early existence. In 1918 a National Technical Sub-Committee was formed for the purpose of selecting and publishing technical papers, particularly those of members. This work continues, together with the production of a series of technical data sheets for use in work.

(5) *From control scheme to pattern bargaining*

One of the methods traditionally employed by the craft and professional associations as a means of exerting bargaining pressure was that of the published list of prices or rates at which members of associations would accept work. If employers refused to pay this rate then members who boycotted the employer were sustained in the interim from the funds of the association. Thus the association was able to exert pressure without resort to open conflict with the employer. A gradual withdrawal of labour, one at a time, is a step towards militant action. Finally comes the direct threat of a strike, and ultimately the collective withdrawal of labour and the seeking of assistance from fellow trade-union-

[1] J. E. Mortimer, *A History of the Association of Engineering and Shipbuilding Draughtsmen* (A.E.S.D., 1960).

ists to prevent the employment of blackleg labour. This model fits the evolution of DATA's collective bargaining policy almost exactly.

Since 1914 a questionnaire has been sent annually to all active members asking for details of earnings and conditions of work. This information is supplemented by the monthly returns of members' earnings made by all Corresponding Members. The results of these surveys are distributed to members in periodic statistical schedules.[1] In this way the DATA (A.E.S.D.) has been able to assess the 'going rate' for draughtsmen. It was decided in 1920 to draw up a scale of minimum rates below which no draughtsman should accept employment. Members are expected to enquire from the union headquarters about the conditions at any office to which they intend to go.

The inadequacy of these attempts at individual job-regulation, supplemented by the Association's policy on training and the setting up of its own employment exchange, became increasingly apparent in the inter-war period. However, the system that was evolved for individual bargaining became the basis upon which the Association was able to build a most effective strategy and organization for collective bargaining in a full-employment economy. In the changed market conditions following the second world war the Association's statistical service became the focus of a network of communications which enabled DATA to become unique in its use of 'pattern bargaining' and in the strategic use of the strike. This form of central 'control', based on the ability to co-ordinate organizational activities, together with a very militant official strategy, made it possible for the Association to boast that it had 'very few unofficial strikes'.[2]

(6) Dilution and its effect on strategy

The outbreak of the second world war again brought all the manifestations of rising demand and government attempts to control the resulting market pressures. Dilution increased membership; most of the new members had been promoted from the shop floor with little or no training and had acquired no 'professional' identity. The earnings differentials *vis-à-vis*

---

[1] The information gathered is now stored in a data bank.
[2] *Royal Commission on Trade Unions and Employers' Associations* Minutes of Evidence No. 36 DATA (H.M.S.O., London 1966).

shop-floor workers were eroded by payment-by-results schemes and overtime payments. Mobility was restricted by government order.

A.E.S.D. tactics, however, were changing. Despite their earlier opposition to nationally negotiated rates, the Executive Council reluctantly entered into talks on this question with the Engineering Employers' Federation in 1939. Strong pressures favouring this development were being generated among delegates to the Representative Council, and finally in 1942 the National Executive was instructed to enter into negotiations for a national minimum rate at age 21; this was achieved in 1944. In 1943 the union sought to secure an agreed minimum scale up to age 25, but this was not to be achieved until 1965.

Although A.E.S.D. had been bargaining at local level for twenty years, these decisions marked a distinct swing towards collective bargaining at a new level—that of a national 'Common Rule' for the whole membership. This development has been said to denote the beginnings of a two-party system in the Association.[1] The influx of dilutees allowed left-wing activists such as Mr G. H. Doughty (Birmingham North) to organize opposition to the piecemeal bargaining policy of the N.E.C. In the pre-war period of widespread unemployment, plant bargaining had not always proved effective. For the 'new men' a policy of 'national advance through national negotiations' was both an alternative market strategy and the basis of a belief in 'trade unionism as class action' which led them to oppose the current 'individualism'.

In 1958 the 'Left' is said to have replaced the 'Right' as a majority on the Executive Commitee. By that time most Divisional Organizers were said to be of the Left. The election of Mr G. H. Doughty as General Secretary in 1952 had coincided with a new aggressiveness in DATA wage claims. During the fourteen years between January 1951 and December 1964, official strikes of DATA members took place at 238 firms. 15,157 members were involved and a total of £910,000 was paid in dispute benefits.[2] During this period the amount paid out in

[1] Graham Wootton, 'Parties in Union Government: The A.E.S.D.', *Political Studies*, Vol. 9, June 1961.
[2] *Royal Commission on Trade Unions and Employers Associations* Minutes of Evidence No. 36 DATA (H.M.S.O., London 1966).

dispute benefits rose to being proportionally higher (in relation to total income) than for any other British union.

For the most part, however, the dialogue within the Association carried out in the Journal and at Conferences during this post-war period was remarkable for dealing with issues ranging from foreign policy to the nuclear deterrent, rather than for its focus on bargaining questions. In the actual carrying out the Union's negotiating policy, the emphasis had been shifted inexorably by full-employment market conditions back to the local plant level. In this movement the ideological debate at national level appears to have had little more effect than that of enabling and perhaps encouraging the use of local bargaining strength. Since the Engineering Employers' Federation has insisted on settling the national claims of manual unions first, DATA has always been able to initiate its national campaign by local bargaining designed to restore the differentials disturbed by the manual workers' settlement.

(7) *Relationships with the Labour Movement*

Although the founders of the A.E.S.D. recognized 'a deep rooted objection to rampant trade unionism' among draughtsmen, the engineering and shipbuilding craft unions have exercised a major influence on the Association's tactics. Many of the recruits who came to the Association by way of shop-floor jobs or apprenticeships were members of the Amalgamated Engineering Union, and it was rational that early discussions should be held with this union, as with the Boilermakers, on jurisdictional matters. In 1918 these extended into terms for a possible merger. Throughout the first world war, workshops adjacent to the drawing office on Clydeside and other centres of A.E.S.D. recruitment had been gripped by the ideas of Guild Socialism. The A.E.S.D. journal was a strong propagator of such views (and has remained so), the culmination of which, if adopted, could be an 'industrial' structure for union organization.

The majority of delegates who voted for affiliation to the TUC in 1918 were, however, not convinced by ideological arguments for affiliation to the Labour Party. Attempts to set up a political fund failed until 1944. At this conference a campaign committee was set up and, in the ballot of members which followed, a two-thirds majority was secured for establishing a political fund. However, only about 37 per cent of members subsequently contracted to pay the political levy, and this number dwindled both absolutely and proportionately during the 1950s, the most militant years in terms of bargaining. The lowest proportion of political contributors is now in the main industrial areas of the Midlands, in which militancy in support of wage claims is particularly high.

These figures suggest that many ordinary members may regard the actions of the Association in a limited and perhaps calculative manner rather than as an instrument of wider social reform.[1] DATA Journal correspondence columns seem to confirm this impression. A former editor has explained that 'at annual conferences and in meetings of its Executive Committee, account has had to be taken of the backward attitudes of some of the members of the Association.'[2]

In the years immediately following the General Strike there was a marked swing towards a more conservative approach in relationships with other unions and in negotiating strategy. The Association did not affiliate to the Confederation of Shipbuilding and Engineering Unions (C.S.E.U.) until 1943, this being 'the logical consequence of the decision of A.E.S.D. to seek national wage agreements'. In 1944 a Joint Consultative Committee of Staff Unions was established under the aegis of the C.S.E.U., in which A.E.S.D. developed strong relationships with the other two technicians' unions, the Association of Scientific Workers and the Association of Supervisory Staffs, Executives and Technicians. The leadership of the Draughtsmen's Association retained a very strong relationship with that of manual unions represented on the Executive Council of the C.S.E.U. Its relations with the other two technicians' unions reached a high point in their joint campaign against the Prices and Incomes Policy of the 1966 Labour Government. Yet neither of the other bodies achieved the same degree of unanimity with the manual representatives in the C.S.E.U. as has been enjoyed by DATA leadership. DATA (A.E.S.D.) has also been much more successful than the others in the co-ordination of its local bargaining strategy with that of blue-collar unions.

[1] Also see Chapter 8.
[2] J. E. Mortimer, op. cit.

(8) *Towards an open union*

Market forces brought a 'dilution' of the draughtsmen's skill, but rapid technological and organizational changes were accompanied by a new and more specialized division of labour in and around the drawing office. A founder member had warned of these forces in 1913 when insisting on high standards of training for entry. In 1942 delegates remained true to the principle of

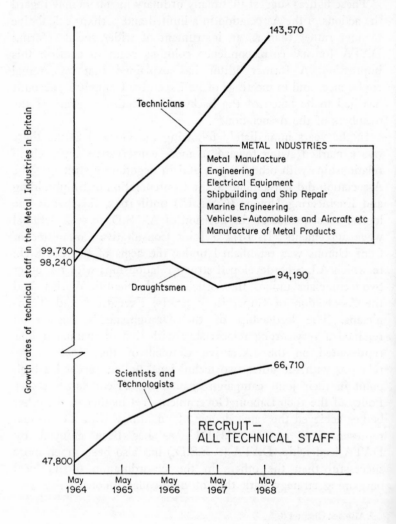

SOURCE: Data Journal, April 1969

'closed membership', and turned down a suggestion that the title be changed to the Draughtsmen's and Allied Technicians' Association and that their field of recruitment should be widened accordingly. In spite of the inclusion of planning engineers among the groups organized by the Association, membership remained relatively static in the early post-war years. Another initiative was taken in 1954, when the growth rate had dropped to a half per cent, but it took the Executive Committee until 1961 to obtain the two-thirds majority required by statute before it could change the Association's title and basis of recruitment. Representational rights were immediately sought with the Engineering Employers Federation (E.E.F.) on behalf of these new groups, but agreement was not reached.

This lack of national success did not prevent DATA from pursuing recognition at the local level. The leadership increasingly encouraged drawing office committees to recruit outside their own occupational group. An example of the argument in support of this policy is reproduced in Figure 1. Frequently a long list of occupations from which recruits are being made is given in DATA *Journal*. The entire range of technicians' jobs covered in the present survey has been included in the list.

Association of Scientific Workers

In 1967 the A.Sc.W. had a book membership of some 22,000, of which about 18,000 were 'paid up'. Roughly 25 per cent of these members were qualified scientists, engineers and technologists, in 1963; the percentage in 1967 is not known, but it seems likely that it fell during the intervening period.[1] The remainder was made up of laboratory and ancillary technicians. These were to be found distributed over engineering and electrical goods, chemicals, and manufacturing industries, with a second important membership in the universities and technical colleges and in the National Health Service.[2]

(1) *Structure and membership*

The 'scientists' find the basis for their organization in their place of work. Unlike the draughtsmen, they do not possess the

[1] K. Prandy, *Professional Employees, A Study of Scientists and Engineers* (Faber and Faber, 1965) estimated that the graduate membership of A.Sc.W. in 1964 was no more than 750.
[2] G. S. Bain, op. cit.

sense of group identity which accompanies a similarity in job
qualification and task performance. In one survey the qualifications of A.Sc.W. members were found to include membership
of the Institutes of Mining and Metallurgy, of Chemical
Engineers and of Electronic and Radio Engineers. Their college
qualifications showed a similar diversity of interest.[1]

The term 'laboratory' was discovered to cover workshops as
varied as the industries served, so that while the members of
A.Sc.W. find themselves employed in large groups working on
similar tasks and to similar standards, these similarities are most
often confined to the firm or to a small sector of the industry.
In addition to the geographical and industrial divisions members
were divided amongst themselves by the status hierarchy of
qualifications and of authority within the work-place.

This heterogeneity of qualifications and conditions of work
leads to a diversity of response in terms of occupational identity
even within the laboratory itself. In practice, A.Sc.W. membership in the firms included in this study covered a wide range
of quality-control and supervisory staff whose association with
laboratory work was somewhat tenuous, but who nevertheless
shared membership of the same local union organization.

Members were drawn together in the branch and on the Area
Committees, to which all work-based groups were entitled to
send at least one representative. Considerable autonomy was
given to Area Committees, which were elected from branches,
but the policy-making body was the Annual Delegate Meeting
or Council, composed of branch delegates. Executive authority
was vested in a committee consisting of (a) national representatives elected at Council, and (b) area representatives appointed
by ballot at Area General Meetings. The resident Treasurer
and Honorary General Secretary were elected by Council, but
the latter office was paralleled by a full-time General Secretary
appointed by the Executive Council in open competition. This
latter post, together with that of Assistant General Secretary,
and those of regional organizers, were 'career' appointments
normally filled outside of the ranks of union membership.
These 'professional organizers' were matched by an Executive
Committee upon which professional scientists—the 'aristocrats'

---

[1] K. Prandy, op. cit.

among lay members[1]—predominated. In many ways the attempt to match the full-time officers with elected lay members resembled that adopted by other high-status groups like DATA.

(2) *Origins*

By 1914 industrial scientists were already being employed in large numbers, particularly in chemicals and in the primary sectors. This recruitment of technologists into industry was accelerated by the first world war. The first moves towards association took place in 1917, against the same background of tight labour-market conditions and of shop-floor militancy, which helped to persuade the Government to recognize the A.E.S.D. in the same year. The title chosen for the organization at the national inaugural meeting of eleven branches, already formed by 1918, was the National Union of Scientific Workers (N.U.Sc.W.).

It is clear that the pioneers, like their forerunners in the drawing office, were ideologically committed to collective action. Indeed, the part-time General Secretary was to become a Labour Member of Parliament in 1924. Among the objects of the Union was 'to regulate the relations between members and employers, and between individual members, or with regard to the classification of members, and for that purpose to impose all such professional conditions as may be considered expedient'.

(3) *Professional association or trade union?*

The means to this end were also contained in the Constitution: 'To set up an employment bureau ... to set up a register of places of employment' and to pay out-of-work benefits. As in the draughtsmen's case, the collective strategy adopted was one designed to exploit the scarcity value of their individual qualifications, rather than to impose a 'common rule' on all those whom it claimed to represent. Yet as a representative body the N.U.Sc.W. obtained early recognition on the technical sections of the Civil Service Whitley Councils. In order to gain adequate representation on the Government-encouraged Joint Industrial Councils in the private sector of employment, the N.U.Sc.W.

---

[1] This title is self-explanatory, but for a fuller analysis of the role of 'aristocrats' in formerly 'closed' unions, see H. A. Turner, op. cit.

D

helped to establish a Federation of Technical and Scientific Workers with other small groups of engineers and chemists. A year later it retreated from the Federation because these latter groups obstructed these collectivist aims.

Membership grew only slowly; from 500 in 1918 to 826 in 1922 (of an estimated 5,000 industrial technologists). Some grievances were successfully taken up by the union, but the nature of membership prevented very effective collective action. A few years later potential members were still not convinced that any 'tangible benefit could accrue from membership'. From the outset membership of the Federation was confined to qualified scientists. Outside the Civil Service neither employers nor their high-status employees were anxious to forgo their personal conduct of relations with each other.

For the Union to succeed it was believed that it had to emulate a professional association and gain some control over the supply of professional scientists. The sheer heterogeneity of scientific qualifications made this a task which the Chartered Engineering Institutes were unable to accomplish over a more limited industrial sector. Yet it was towards this course that the union was steered. An early partnership with the British Association of Chemists foundered after two years. A later attempt to control other professional places of employment through an alliance with the newly established Institution of Professional Civil Servants, together with the Association of University Teachers and the rival British Science Guild, also came to nothing. Meanwhile the union had adopted a model for minimum scales and conditions which included equal pay for women. Their objective reflected an allegiance to ideals rather than an appreciation of the reality of industrial relations in the 1920s.

All of this programme proved futile. Despite an initial aim of the Union 'to advance the interests of science, pure and applied, as an essential element of national life', potential members in still small factories and laboratories remained uninterested in being pioneers of what was manifestly a trade union. A number of eminent 'committed' scientists and employers appeared among the members, but this had little impact. In 1922, the Executive announced that of the joint aims of advancing the interests of science and of improving the conditions of scientific workers the former took first place. *'Scientists should not be regarded as employees*: they have special responsibilities to the nation.'

In keeping with the policy of other groups with which the N.U.Sc.W. had been co-operating, Union policy 'now needed to represent itself as an organization like the British Medical Association', co-operating with professional institutes to ensure that only 'those professionally competent should carry out the practice of science'.[1]

In 1925 membership fell and it was finally decided to de-register as a trade union and to change the aims, as follows:

(a) to promote the development of science in all aspects and to maintain the honour and interests of the scientific profession, and
(b) to secure that the practice of science for remuneration shall be restricted *by law* to persons possessing adequate qualifications and to co-operate with those bodies empowered to grant certificates of qualification in order that high standards of professional competence should be established and maintained.

The name was also changed to the Association of Scientific Workers.

There followed an immediate influx of some 610 new members, which, though raising membership some 77 per cent, still left the 20,000 industrial scientists relatively little affected. By this time the economic slump had set in. The real earnings of scientists were increasing in a period of falling prices, but the disincentives to any form of collective action were very real. By 1930 membership had fallen to 1,000. The losses were particularly concentrated in the industrial sector, and the Association appears to have been kept going by groups in academic laboratories, who were at this time subject to Government-imposed cuts in salaries.

The exiguous membership of the Association was severely hit by events following the general strike. Like the draughtsmen, the A.Sc.W. established a working arrangement with a civil service association after the 1927 Trades Disputes Act: but in its arrangements with the Institution of Professional Civil Servants it lost its negotiating role entirely. During the decade 1925–35, the A.Sc.W. attempted to act as a 'protective associa-

---

[1] For an interesting example of similar 'premature' unionization of professional groups see Archie Kleingartner, 'The organization of White Collar Workers', *British Journal of Industrial Relations*, Vol. VII, No. 1 (March 1968).

tion'[1] and failed. But under its new constitution it carried out certain 'learned-society' functions in the field of industrial education, in which it was successful in bringing about innovations.

(4) 'Dilution' and its effect on strategy

In 1937 the Executive Report reminded members of the 'principal object to advance the interests of scientific *workers*', and in the following year industrial branches were formed. Early in 1940 the series of local industrial conferences began a shift in the balance of members, and recruitment in manufacturing industry improved as a result. In November 1940, the rules were altered to enable unqualified assistants 'working in a technical capacity under the supervision of persons so qualified' to be recruited. The Association was then in the position to recruit the technicians who had entered the science-based industries of the late thirties. The same Council meeting decided to re-register as a trade union.

(5) Growth and recognition

In 1940 membership increased by 25 per cent, in 1941 by 125 per cent, in 1942 by 100 per cent, and in 1943 by 67 per cent, when it reached 14,010. The growth in membership forced rapid changes in organization. The Association's new strength enabled it to shift the basis of organization to the place of work. The first full-time Industrial Organizer was appointed to undertake recruiting and negotiations. Such negotiations became more and more frequent and successful at plant and company level, with recruitment spreading into several industrial sectors. Nevertheless in 1942 the strength of the A.Sc.W. was spread thinly, and it had no more than 3,400 members in engineering. A request for a procedural agreement' made to the Engineering Employers' Federation was refused. The grounds for refusal were that 'a close personal relationship exists between managements and their technical staffs and conditions of employment could be determined only through this personal relationship'.[2]

[1] This term is used in the sense in which it appears in G. Millerson, *The Qualifying Associations* (Routledge and Kegan Paul, London 1964).

[2] G. S. Bain, op. cit. Similar opinions were reiterated by some managers in the course of this study.

It became clear, at a meeting arranged through the offices of the Ministry of Labour in 1942, that the major objection of the Engineering Employers' Federation to recognition of the A.Sc.W. was that many of its members occupied executive or managerial posts in industry. This was to provide a test of the real interests of the Association. In the event the Association declared itself to be only interested in technician grades of staff, and not managerial grades or technologists. Since the London Employers' Association was ready to recognize the Association, the Engineering Employers' Federation was persuaded to enter into negotiations nationally. These began in February 1943 but dragged on over the next fifteen months. Eventually the A.Sc.W. reluctantly agreed not to include technologists in its negotiations, and they also unwillingly agreed to the employers' right to alter salaries before exhaustion of procedures. This limited procedural status sealed the identity of the A.Sc.W. as a technicians' union on similar lines to that of the draughtsmen.

(6) *Relationships with the Labour Movement*

Confirmation in its trade-union ideology was marked by affiliation to the TUC in 1942. At plant level it was already developing a common front with ASSET and A.E.S.D., which led up to the decision to affiliate to the Confederation of Shipbuilding and Engineering Unions and to participate in the Joint Consultative Committee for Staff Unions formed within the industry in 1944.

The setting up of a political fund seems to have been even more controversial in the A.Sc.W. than was the case in DATA. The N.U.Sc.W. had affiliated to the Labour Party and had supported the General Strike—with results to the membership which were regarded by some as disastrous. At various times after 1940 the question of reintroducing an official political policy was raised. In setting up its political fund in 1946 the Association provided a compromise solution. A political levy of 2s. 6d. per year was to be made, but there was no guarantee of support for, or affiliation to, any political party. In fact, the political fund turned out to be so weakly supported that its existence would appear to have been almost irrelevant. In the decade 1955 to 1965 the fund never held more than £92, and in 1959 contained only £45 4s. 3d.

Nevertheless many lay and full-time officers held left-wing views and were in favour of trade unions playing an active part in national politics. Throughout the 1940s and 1950s references were made by members to the political image projected by the A.Sc.W., not every active member being happy with the political opinions expressed by the Union's leadership.

At the 1948 Annual Council, one delegate accused the Executive of a Communist bias, and at the 1949 meeting it was stated: 'It is evident that branches are perturbed that the actions of some officers and members have given support to the Communist Party.' As an issue, however, the political views of the leadership faded in the 1960s, despite descriptions of Communist affiliations among them appearing in the press.

(7) *Towards an 'open' union and organizational inaadequacy*

It is clear that, as in DATA, the leaders and local activists in the A.Sc.W. were motivated by a collectivist ideology much stronger than and even in many cases foreign to that accepted by card-holding members. In DATA's case the homogeneity of the labour-market and work situation maintained a sense of occupational 'solidarity' which provided a spring-board for action on collectively felt or expressed grievances. A.Sc.W. had the problem of resolving difficulties which were unique to the work situations of their multifarious groups of members. It makes a clear-cut distinction between two groups of members: the qualified and more senior on the one hand and the less qualified members on the other, who often only marginally match the titular status implied by 'scientific worker'. This is not to say the ideological conflict described earlier followed that of the dichotomy in the membership. From the slight evidence available it appears that A.Sc.W. had always been led by an Executive Committee with a substantial number of well-qualified scientists and technologists who could maintain a professional stance whilst holding left-wing political views. The left-wing position of the Association in its wider social views appears to have been introduced by the 'dilutees' recruited after 1940. These less-qualified members made up the majority in January 1968, when the Association amalgamated with ASSET. Yet the first General Secretary became a Labour Party M.P. and long before the re-registration as a trade union the 'learned society' activities of the Association had moved sharply to the left in terms of the subjects discussed.

In spite of the lead given by the politically committed leaders, members have always reacted in terms of their labour-market situation. The founders established a 'union', yet when the 'professional' strategy of closed entry and a limited focus of membership appeared likely to be successful, this was the strategy reluctantly adopted by the leadership after four successive years of Council debates. On the other hand their strongly expressed political commitment did not help to retain the members brought into the Association during the years of the second world war. Membership fell from 17,211 in 1946 to 11,366 in 1955, and remained fairly static until 1961–62. In the early post-war years some fall-off could be accounted for in terms of demobilization of scientists who had been working in Government laboratories. But the long period of wage restraint from 1947 to 1950, in which A.Sc.W. leaders were reluctant accessories, must also have been effective in dampening enthusiasm.

However, over the next five years up to 1965 book membership reached 22,000. These were the years of unprecedented aggregate growth in both technical and scientific occupations. Between 1951 and 1961 numbers of laboratory technicians grew from 69,000 to 96,000, many being employed in large units under 'industrial' conditions. The failure of A.Sc.W. to grow apace in a period in which the fears and frustrations which accompanied these rapid changes in market and working conditions were constantly leading to spontaneous outbreaks of local collective action among technicians can only be attributed to leadership and organizational failures. The first major official strike is believed to have occured as recently as 1962,[1] and it is clear that the left-wing stance of the leadership did not automatically lead to a militant bargaining strategy.

From case histories in the present study it seems clear that, as with DATA, local leadership was heavily dependent upon lay membership rather than full-time officials. Local officers did not, however, have access to a network of communications such as that developed by DATA. Neither did they deal with the uniformity of substantive issues that are the hallmark of occupational homogeneity. The plant committee of A.Sc.W. had to

---

[1] This is put forward by K. Prandy, op. cit., on the basis of available A.Sc.W. records

encompass the wide range of levels of remuneration and conditions within the limits of their membership. In these circumstances a great amount of time and personal commitment was
required of the local leaders. In the one plant in which A.Sc.W.
was particularly strong, leadership at local level, as at national
level, rested with qualified technologists. Such men were already
strongly committed to their work role, and some to professional
institutions. The heavy demands of negotiating for several
grades of laboratory assistant, test-gear inspectors, test engineers,
etc., required more of them than just ideological commitment;
it required the expertise of a professional negotiator.

In this respect A.Sc.W. gained greatly from affiliation to the
Confederation of Shipbuilding and Engineering Unions and its
association with DATA and ASSET, on the Joint Consultative
Council, and with other unions on Joint Industrial Councils
and Whitley Councils. At local level, however, its apparent
inability to match up to the organizational challenges of an expanding labour market seemed to underline the need for professional 'trade union managers', a need which followed from the
heterogeneity of its membership. However able, intelligent and
strongly motivated were the original activists, these factors were
no substitute for effective organizations. Indeed, it may be that
the existence of a number of strong 'professional' identities
within an organization which attempts to adopt the functions of
a 'protective association' necessitates even more exacting services
than those usually provided by a trade union.[1]

Association of Supervisory Staffs, Executives and Technicians

In December 1967 ASSET had a membership of 58,500.
Approximately half of this number were foremen; the other
half was composed of a wide range of members in technical
occupations. These were mostly contained within two internally
defined grades of membership; a third grade admitted 'persons
acting in a full-time capacity in positions of higher responsibility than Supervisory and Technical Grades'. Membership of this
third managerial grade was relatively insubstantial at that time.

---

[1] As a contrast, see for example the comparative success in this field of 'professional workers' working within one sector achieved by NALGO.

(1) *Structure*

The basis of organization was (and in the A.S.T.M.S. remains) the workplace-based group, each of which elected an accredited representative for the purpose of communication and negotiation. Where possible the group made up a branch, or, if the plant was too small to allow this, it was attached to an area branch. There were eleven District Councils, each of which was also an electoral district for balloting to the N.E.C.

The organizational structure of ASSET was only superficially similar to that of the other two unions. The rationale of ASSET's organization was to provide a professional service to groups who were too small, too localized, or generally too devoid of formal qualifications or other means whereby they could 'close' entrance to their occupation. Mobility between jobs did not offer the same career prospects to these foremen technicians as it traditionally did to the draughtsman or would-be professional. As the union leaders expressed it themselves 'the main benefit —and purpose—of ASSET membership is the industrial counselling and negotiating representation that we provide.'[1]

This service was given by a staff of thirteen senior officers— ten of them Industrial Officers—together with the staff attached to each of the District Offices. In 1966 ASSET announced its intention of increasing its field force by 50 per cent over a three-year period; it also decided to provide legal services as a vital adjunct. Whenever strategically important negotiations were entered into, or when a crisis involving ASSET in exposure to the mass media occurred, it was the practice for the General Secretary to become the union spokesman.

Thus, as in the large general or 'new' manual union, the style and practice of organization descended from the top, and its effectiveness depended upon the expertise supplied by its corps of professional Trade Union 'counsellors'. The Annual Delegate Conference was such that its business could be concluded in two days. Moreover, the emphasis in the Conference agenda was always placed on the Executive Report rather than upon Branch motions. Its Journal was largely written by the General Secretary and the National Officers, and contained no membership correspondence.

---

[1] *Royal Commission on Trade Unions and Employers' Associations*, Minutes of Evidence No. 53, ASSET (H.M.S.O., London 1967).

This does not imply that the members' wishes went unheard. On the contrary, the existence of local and national 'counsellors' saw to it that grievances were investigated promptly. It means, rather that the nature of the organization was one of specifically servicing the interests of members in their work role, rather than attempting to obtain a 'moral'[1] commitment of the kind demanded by the old-style trade-union activist of his 'card-holding' members. The ideology of protest became a tool of organization rather than springing from spontaneous collective action.

(2) *Origins*

Like the A.Sc.W., the Association of Supervisory Staffs, Executive and Technicians really assumed its modern character and structure during the second world war. The basis for this development existed in the organization of the National Foremen's Association, which had been supported by a small group of foremen members since it was first formed in 1917. No fewer than twenty-two foremen's associations were organized between 1916 and 1918. The earlier existence of friendly societies for supervisors and foremen dating back into the nineteenth century had already caused employers to respond through the formation of the National Foremen's and Staff Mutual Benefit Society in 1899.

Although prevented by the nature of their job from the close contact with their peers which were enjoyed by draughtsmen and industrial scientists, foremen were perhaps subjected to far greater stress throughout the whole of this period of rapid change. In many instances they would be bearing a heavy load of responsibility in effecting the changes, and yet, like the men they supervised, they were extremely dependent upon a single employer for the maintenance of their status, as for their superior position in the labour market. This dependency would be accentuated during the periods of rising shop-floor earnings, when they might find they were being paid less than the men they supervised.

The major area of recruitment of the National Foremen's

---

[1] The term is used here in the sense in which it is defined by A. Etzioni, *A Comparative Analysis of Complex Organisations* (Free Press of Glencoe, New York 1961).

Association was within the large railway workshops, while yet other pockets of membership were recruited in the engineering industry. The foremen who joined were often ex-members of craft unions and 'were spurred on . . . by the recognition of the rail unions and their subsequent gains.'[1]

Membership spurted to nearly 3,000 in 1920, but thereafter dwindled during most of the inter-war period. The National Foremen's Association had taken a decision to affiliate to the TUC in 1919 and was clearly seen as a trade union. On the other hand 'the great mass of the foremen considered their interests would be best served by their lining up with their managements, joining company promoted internal staff associations, etc.'[2]

(3) *Recognition and growth*

The identification of the National Foremen's Association as a trade union was to stand it in good stead when in 1942 it sought recognition as a representative body within the engineering industry. Not having the strength of membership to negotiate effectively with the Engineering Employers' Federation, its leaders decided to solicit the aid of the T.U.C. and Parliamentary politicians.

The Association won an early victory for recognition in 1941 in the Northern Ireland engineering firms Harland and Wolff and Short and Harland. After approaching the Government for arbitration on their claims, the firms were persuaded to conduct a referendum among foremen in which the Association subsequently gained the support of 70 per cent of these employees. The Engineering Employers' Federation remained intransigent in their argument that 'as foremen are fundamentally part of management, their conditions of employment cannot properly be the subject of negotiations by a trade union organization.'[3]

The following year the Association made the first of its innovating gestures in bargaining tactics for which it has subsequently become famous. Its officers submitted a claim for salary increases for foremen employed at a plant of the General Electric Company in Birmingham. Upon the refusal of management to

[1] Minutes of *Royal Comission on Trade Unions and Employers' Associations* Evidence. op. cit.

[2] T. W. Again, ASSET General Secretary, 'A Call to Foremen and Engineering Technicians', September 1945.

[3] Quoted by G. S. Bain, from a report of an observer to the Engineering Employers' Federation.

negotiate, the union submitted a separate claim for each of its 280 members to the wartime National Arbitration Tribunal. At a mass meeting staged in London, Sir Stafford Cripps, Minister of Aircraft Production, announced that he had invited the Association to set up a committee to advise the Ministry on technical production matters. The General Secretary responded by saying that with recognition from so 'important a Ministry they need not worry unduly about other quarters.'[1] Within a week the Association was invited to talks by the E.E.F., and an agreement was signed one year later. Once more the Government was instrumental in obtaining employer recognition.

Recognition was, however, only partial and confined to those firms in which the union had majority membership among the grades for which it catered. These terms had been suggested by the General Secretary of the T.U.C., Sir Walter Citrine, who had previously approached the E.E.F. on two occasions, informing them of the 'fullest support of the T.U.C. for the Association'. The influence of this support cannot be discounted, but the impact of the Association's tactics and allies in Parliament appears to have been decisive at this time. Especially important was the association with Mr Ian Mikardo, left-wing Labour Member of Parliament from 1945 and later successful business management consultant; up to the present time this association has given ASSET an invaluable source of political influence and support.

A change in title in 1942 to the Association of Supervisory Staff and Engineering Technicians, and then to the Association of Supervisory Staffs Executives and Technicians, indicated not just an 'opening up' of eligible membership, but also a 'bidding up' into higher status groups within the plants in which the Association organized. Membership grew by 62 per cent during 1943 to 10,820 and showed a further advance to 15,809 in 1948. However, this increase was described by its leaders as 'unreal and wasted away with [post-war] redundancy'. Membership declined drastically in 1949–50 to around 12,000, and only slowly began to climb to a point at which it overtook its immediate post-war level.

The organization was based upon several industrial sectors, the major ones in 1946 being airways, railways, engineering, metals, electricity, gas and chemicals. At this stage ASSET still

[1] G. S. Bain, op. cit.

appeared to be concentrating its efforts on foremen despite its change in title. Its journals and reports, though conveying a Socialist message in articles and editorials, mainly comprised reports on key negotiations and settlements. Its recruiting literature argued that the status accorded to the 'rank and file of the technical, scientific and managerial grades' was inadequate, but also urged those in these grades towards 'the final grand objective of the British Labour Movement—the completely planned Socialist society'.[1]

(4) *Relations with the Labour Movement*

It was not until 1945–46 that a ballot of members was taken which showed a majority in favour of setting up a political fund. Affiliation to the Labour Party was applied for, and ASSET's first delegates attended the Labour Party Conference in 1948. Affiliated membership at that date, and for the next few years, stood at about 25 per cent. The political fund is not mentioned in the Annual Report again until 1959–60, when it was suggested that 'too few members were contributing to the political funds'. In 1960–61, however, the political levy contracting-out form was amended (within legal limits), to make contracting-out less easy. Over the ten years 1955–65, according to the Registrar of Trade Unions' file, ASSET's political fund moved upwards from £482 to £4,854.

The rate of growth in the number of members paying the political levy increased rapidly over the years 1961 to 1965; this has been attributed by the General Secretary to 'a programme of political education' launched by the Executive Committee.

During the whole of this post-war period the leadership has used the platform provided by the T.U.C. and Labour Party to put forward the left-wing socialist ideology. On many occasions this view has been expressed in relation to a bargaining issue such as the Association's opposition to incomes restraint in 1948–49, 1961 and from 1964 onwards. However, on other issues ASSET's opposition has had to be expressed in somewhat ambivalent terms, for example, on such matters as defence expenditure, when members' jobs were to be considered. Or, again,

[1] 'Managers, Supervisors, Technicians: Your future and status' (ASSET, 1946).

while DATA called for a general strike against the Government measures to reform industrial relations in 1969, the leaders of A.S.T.M.S. were by no means so clear about the nature of the protest in which their members should (or would) participate. This ambivalence in matters involving their members' commitment to the views of their leaders can only be explained in terms of the service being provided by the Association to a group with only a relatively narrow form of loyalty to the organization.

ASSET attempted to provide a counselling service for the British middle manager who 'was not going to defend himself'.[1] To other technicians' unions concerned with the maintenance of their white-collar members' job opportunities, ASSET's position was a familiar and operationally acceptable one. Hence ASSET enjoyed a good working relationship with these unions on the Joint Consultative Council in engineering and on similar bodies in their other more recent fields of recruitment—transport, chemicals, metals and gas. Joint executive meetings of the Draughtsmen and Allied Technicians' Association, Association of Scientific Workers, Society of Technical Civil Servants and Association of Cinematograph, Television and Allied Technicians were held from time to time. ASSET also laid great stress on its policy in relation to jurisdictional and demarcation issues, 'seeking to obviate any possible friction in areas where there might otherwise be competition for membership'. It prided itself on consistently observing the Bridlington Rules of the T.U.C., and upon the jurisdictional arrangements it had made with other unions outside the T.U.C.

Yet despite this formal *modus vivendi* ASSET never appears to have been wholly accepted by the movement in the way that DATA has been accepted, or even more recent arrivals on the General Council of the T.U.C., such as NALGO. The failure of the General Secretary of ASSET (by then A.S.T.M.S.), Mr Clive Jenkins, to secure the seat on the Council created for the Professional, Scientific and Technical sector of T.U.C. membership in 1968 was not unexpected. The size of the majority, 6½ million on a card vote, by which the draughtsmen's General Secretary, George Doughty, was elected was evidence of a widespread antipathy towards A.S.T.M.S. and towards its methods. The acrimonious exchange at the 1967 annual conference of the Confederation of Shipbuilding and Engineering Unions, when

[1] *Sunday Times*, 1 December 1968.

Mr Jenkins had led his union's delegation out of the hall during a debate on incomes policy after the Chairman had refused his 'point of order' and instructed him to 'hop it'[1] was symptomatic of the antagonism aroused by A.S.T.M.S. and the style of its leadership.

(5) Strategy for growth

In part this antipathy sprang from the forthright statements of the ASSET leader, Mr Jenkins, on the 'realities' of the labour market and the successes of his organization in meeting them, successes which were by implication contrasted with the achievements of other unions. Writing on white-collar employees in 1963, he said 'quite decisively ... they are declining to join the bigger traditional manual unions and insisting upon special (and status conscious) bargaining agents ... the failure of certain militant old unions to trace and claim members who became "gaffer's men" in all sorts of professional and technical jobs may be seen as a source of their growing impotence'.[2] The contrast between organizational strategies of the 'old' manual unions and the 'new' as represented by ASSET was and remains part of the successful recruiting strategy of A.S.T.M.S. leaders. During the last ten years of its separate existence ASSET membership increased by nearly 200 per cent, from 20,172 to 58,500. Mr Jenkins explained the reasons for this growth as being centred in a 'specific change in organizational policy, policy which was set in train ... in 1958 ... [and] which called for an end to the indiscriminate, catch-as-catch-can organizing of the past.'[3]

The new directions in ASSET's organizing policy were based on the quite novel assumption that given 'the movements in industry towards monopoly, oligopoly, takeover and rationalization ... it would only be possible to build up union financial strength and membership rapidly and efficiently by organizing the big employers'.[4] It was argued that organizing small firms and obtaining settlements in them would have minimal effect on national wage patterns compared with successes in the larger industrial concerns.

---

[1] *Daily Telegraph*, June 1967
[2] *New Society*, 30 May 1963
[3] 'Tiger in a White Collar?', *Penguin Survey of Business and Industry*, 1965.
[4] ibid.

This 'combines' policy, as it has since come to be known, necessitated a number of changes in the organizational structure of the Association. Whereas previously it had been organized on a purely district basis, the new policy required an organizer to deal specifically with one big firm or firms. It is his responsibility 'to study it intimately so that he is aware of the national policy as well as the local problems'.[1] ASSET purchased a token number of shares in key concerns, and the 'specialist official'— as shareholder—'goes to the annual meeting and, if necessary speaks up . . . about issues which might be difficult to express as a trade union officer in factory collective bargaining'.[2] In order to co-ordinate his combines policy, he further had the responsibility of 'organizing an annual conference of delegates from all the factories (say) in Hawker Siddeley or General Electric'.[3] The result of these conferences was intended to be a 'common policy for levelling upwards in the coming year'.[4]

Although policy changes may be highly important in accounting for the success of a particular organization, Clive Jenkins did not attribute ASSET's advances solely to these changes. He echoed some of the statements made by DATA officials in saying that the attempt to enforce wage restraint was an important contribution to white collar unions' recruiting appeal. Of the 1961–62 pay pause he said

> The active, the clever and the frustrated who had been 'paused' looked around for a vehicle to carry their resentments; they found it, often to their surprise, in the unions that had been set up and were standing bonily waiting for this fleshing-out to happen.[5]

ASSET's opposition to the Government's incomes policy— and the lengths to which they have carried this opposition in the face of criticism from the Labour Movement itself— indicated that they placed great faith in this kind of action as a means of recruitment. Confident statements in the ASSET journal implied that the policy was having the desired effect.

(6) *Towards a 'middle-class consciousness'*

Mr Jenkins has described the way in which he sees the social

[1] 'Tiger in a White Collar',
[2] ibid. p. 58
[3] ibid. pp. 58–9
[4] ibid.
[5] ibid.

processes working within the labour market: 'there is a grow-ing sense of identity with people in the same occupation which extends to other units in nationally distributed organizations.'[1] This he illustrates with examples of frustrated aspirations and with the breakdown of individual treatment within the large firm. The envy felt for the local bargaining power of shop-floor workers under incomes policy is merely a catalyst in this process. If this thesis holds good then the frequent and widespread redundancies among middle management and technicians which have followed from Government-encouraged mergers between employing firms in the 1960s must have served to emphasize the insecurity of this group in a new and exposed market posi-tion.[2] In these situations ASSET has consistently acted in a manner which gave maximum publicity to the plight of its members, but placed a minimum reliance on the collective action of its members.

While DATA place great emphasis on militancy and are prepared to spend substantial amounts on dispute pay, ASSET seem to have put less reliance on industrial action. Levels of dispute expenditure are not necessarily the best index of militancy; but they do in ASSET's case indicate how much of a newcomer to the high dispute-spending arena the Association is. In 1955 ASSET spent only £15 on dispute pay. In 1965 this reached a record level of £63,918.75, but on average its subse-quent expenditure (including that of the A.S.T.M.S.) has been no more than a fifth of the latter sum.

As might be expected, many of the more recent ASSET (and A.S.T.M.S.) strikes related to the issues of recognition. Though ASSET prided itself on the lack of unofficial action within the current membership, it was apparent from the results of this study that collective action of an entirely spontaneous kind had often preceded the introduction of ASSET into a situation. From such actions as these and from the ideology of protest used by the ASSET leadership to express these frustrations the Asso-ciation had acquired its reputation for militancy. Without the 'bare bones of unionism waiting to be filled out' it seems

[1] *New Society* 30 May 1963
[2] ' "Takeovers" in the period 1960–64 had increased by three times on the previous period of four years but by 1966–68 they had increased by a further three times.' *The Draughtsman*, October 1968.

doubtful whether this protest would have reached national proportions over such a short period.

(7) *Towards an institutional convergence?*

In many respects the evolutionary paths of these three institutions are much as might be predicted in the Webbs' historical categorization of unions and in Professor Turner's more recent treatment of union growth and development. The initial movement towards a more open recruitment policy was brought about by increases in the level of demand. The spread of recruitment needs to be treated separately. This took place along a largely horizontal path into allied occupations in the case of DATA, vertically downwards among laboratory assistants, etc., in A.Sc.W.'s case, and vertically upwards into the ranks of technologists and managers in that of ASSET. Changes in the elasticity of demand brought about by technological and organizational innovation indicate a fresh division of labour in the place of work. Draughtsmen are now less responsible for the design function than ever before; meanwhile, the ratio of other technicians to industrial technologists has steadily increased since 1937, each new division eroding the central work-role of the original occupational group.

But this rather deterministic model is capable of being modified in two important respects. As Professor Bain has pointed out, the attempts made to recruit membership by a union may eventually depend upon their success in obtaining the employers' recognition and even endorsement of their actions. In other words, the union has to be able to offer a representative service to prospective members (the 'skeletal form'). Secondly, and perhaps more importantly, the existing union members must be satisfied, or in some way reconciled to sharing their occupational representation with employees outside their occupation. The importance of this internal political process should not be underestimated; despite the left-wing stance of A.Sc.W. leaders there can be little doubt that their view of the closed nature of their organization was such as to cause more than a little hesitation at the prospect of merger with ASSET. Or again in the case of DATA, fifty years after draughtsmen's leaders had their first merger talks with leaders of the manual engineering union, their members were finally persuaded to give their consent to this industrial marriage in a referendum taken in April 1970. Yet, as Professor Turner has pointed out, the evolution towards openness in recruitment policies leads to increasing rivalry for membership among the same occupations.[1]

(8) *Convergence and competition*

In October 1968 *The Draughtsman* (DATA's journal) published an article entitled 'Can Data Survive?' in which the Chairman of the Recruitment and Membership Committee concluded that 'only an enlarged and organized membership will be able to survive in an industrial society controlled by giant monopolies.' But the threat described in the article was not that of giant monopolies but of 'the A.E.F. which in particular is determined to maintain its strength by catering for the growing ranks of white-collar technicians'.

The context in which this article was written was strongly influenced by the amalgamation of ASSET and the A.Sc.W. in January 1968. The two unions brought together some 70,000 members under the name of the Association of Scientific, Technical and Managerial Staffs, thus rivalling the position of DATA in the sphere of technician organization. Over the preceding period the recruitment areas of ASSET, the A.Sc.W. and DATA had grown dangerously close together. Some occupational groups, such as planning and estimating technicians, and work study and systems analysts, were to be found in all three unions. Despite their close working relationship, disputes sometimes broke out at local level. Such a clash occurred between A.Sc.W. representatives and DATA members during a dispute at Plessey.[2] Evidence of antipathy between established A.Sc.W. branches and 'newcomers' among ASSET membership came to light in the case studies made for this research.

In all of this, the relationship between other members of the C.S.E.U., especially the A.E.F. and DATA, stood in marked contrast to their 'coolness' towards ASSET. As has been pointed out earlier, this reservation was in some respects induced by the

---

[1] Professor H. A. Turner and the Webbs were aware of the constraints upon union growth. However, it does not appear sufficient to suggest that the reason for union expansion is simply that 'nature abhors a vacuum' as Turner suggests and it seems probable that the *internal* sources of growth are insufficiently examined by these authors.

[2] *The Draughtsman*, January 1968 and April 1968.

statements of ASSET. Although ASSET has encouraged dual membership among promotees to supervisory grades 'in order to maintain good relations with craft unions', in 1966 it announced: 'It is now our feeling that the union to which an employee should belong is the union which holds the negotiating rights for his class of employment and we have been taking slow and prudent steps to this end.'[1] This new, tougher line led to its first major conflict with the Iron and Steel Trades Confederation, an industrial union in which white-collar workers made up 11,750 of their 105,400 members in the iron and steel industry. In July 1967, just before the day assigned for vesting the newly nationalized British Steel Corporation, an agreement between the Corporation and the T.U.C. was announced respecting the future national recognition of unions within the industry. In this agreement recognition was confined to five unions plus the craft unions, the biggest of which were the A.E.F. and E.T.U. These were the longest-established unions and included the Iron and Steel Trades Confederation (I.S.T.C.), together with other largely manual unions with some white-collar membership.

The Clerical and Administrative Workers' Union (C.A.W.U.), ASSET, the A.Sc.W. and DATA had been established in the Steel Industry for some years, and their protests at exclusion from recognition were aired at the 1967 T.U.C. Conference and in the headquarters of the Corporation over the following twelve months. In June 1968 C.A.W.U. and A.S.T.M.S. members took part in one-day strikes, and in August a Court of Inquiry was set up to investigate the reasons for the dispute.

The Court determined that C.A.W.U. and A.S.T.M.S. should receive national recognition according to their criteria for 'a selection of the unions most suitable for the white-collar workers'. The existing steel unions, in their capacity as members of the Steel Committee of the T.U.C., rejected the Court's findings. Eventually in December the Corporation announced limited local recognition for the A.S.T.M.S. and the C.A.W.U. The constituent organizations on the T.U.C. Steel Committees responded by instructing its members 'to take orders only from

[1] *Royal Commission on Trade Unions and Employers' Associations*, Minutes of Evidence No. 53, ASSET (H.M.S.O., London 1967).

staff who are members of the appropriate organization'—the appropriate union for white-collar staff being the I.S.T.C.[1]

The dispute is by no means settled. It has brought the A.S.T.M.S. into head-on collision with the T.U.C. General Purposes Committee, whose embarrassing inability to handle the jurisdictional issue was made plain at a time when many external critics were pressing for trade-union reform. But possibly more important for the developing structure of the technicians' unions was the means whereby DATA achieved recognition, when in March 1968 the decision of the Craft Co-ordinating Committee in the Steel Industry to accept DATA into membership was ratified by the T.U.C. This decision strengthened the bargaining position of the other two white-collar unions, but it was also announced to DATA membership as evidence of the Association's close relationship with the A.E.F.[2]

In October of that year an example of the A.E.F.'s relationship with the A.S.T.M.S. was revealed in a demarcation dispute at the Girling Brakes factory at Bromborough. A.E.F. shop-stewards informed management that their members would no longer accept instructions from certain charge-hands who were members of the A.S.T.M.S. The incident did not appear to have been a very typical one at the time, but it was later to appear crucial in determining the future strategy and growth paths of technician unions.

(9) *Amalgamation and changes in character*

Throughout the period of these incidents, talks had been proceeding between DATA and the A.E.F. to establish 'an agreed basis for amalgamation'.[3] These talks had commenced after the Executive Council of DATA reported to the 1967 Representative Council a failure to reach agreement with the A.Sc.W. in merger talks during the previous year. In 1966 a report suggesting 'the establishment of one organization as a union for all technical staff' had been accepted by the Executive Council of DATA. The General Secretary of the DATA commenced talks between his union and the A.Sc.W. that could

[1] *Report of Court of Enquiry under Lord Pearson into a dispute between the British Steel Corporation and certain employees*, Cmnd 3754 (H.M.S.O., London 1968).
[2] Annual Conference Report in *The Draughtsman*, June 1968.
[3] ibid.

pave the way to a wider merger with ASSET and others. He said that '... in the absence of that we ourselves should build up our own organization in order to become the most important single technicians' union ...'

The A.Sc.W. were also enthusiastic, seeing the resulting scope for recruitment as being 'almost limitless'.[1] Some members were reported as having argued that the union should amalgamate with ASSET, but the A.Sc.W. executive (on which the better-qualified 'professional' interest was in a majority) saw too many difficulties in the short run. At that stage informal talks had in fact been going on for some time between the three unions, but at the 1967 Representative Council, the General Secretary, Mr Doughty, reported failure to agree on possible terms for amalgamation with either of the other parties. The reasons given were that the A.Sc.W. was not an engineering-based union having many members in public and academic establishments, and Mr Doughty later maintained that the 1966 Conference decision 'was for a union based on engineering, shipbuilding and related service industries'.[2]

This interpretation was disputed throughout a two-year dialogue in the correspondence columns of *The Draughtsman*. The counter-arguments were two-fold. The drawing office remained a basis for the career development of increasing numbers of highly qualified young entrants, and their interests and aspirations were not those of a manual union. Secondly, the 200,000 technicians in posts, while large enough for DATA's future development, would remain a minority in the A.E.F. for many years.[3] Nevertheless, the weight of the unions' editorial and educational effort over the period reflected the leadership's attachment to industrial unionism as opposed to a craft base.[4] At the 1969 Conference, delegates voted 'overwhelmingly' to amalgamate with the A.E.F.[5] Under the 1964 Trade Union Amalgamation Act a national poll of membership had to be taken and had to show a majority of members in favour in order that the Executive could go ahead with the merger plans. In April 1970, when such a poll was taken, two-thirds of the

[1] *The Guardian*, 9 May 1968.
[2] *DATA Journal*, June 1968.
[3] Précis of article in *DATA Journal*, November 1967.
[4] See, for example, *DATA Journal*, August 1968.
[5] *The Guardian*, 2 May 1969.

then current membership registered a vote and three-quarters of this number voted in favour of the merger.

The leadership of DATA appears to have been attracted by the concept of an engineering industrial union, with DATA forming the technicians' section, at least as long ago as June 1967, but the factor which appears to have been decisive in uniting the Executive Council behind its General Secretary was the election of Hugh Scanlon to the Presidency of the A.E.F. and the accompanying influx of left-wing candidates to the National Executive Committee of that union in 1968. Until that time, despite extensive local co-operation between the two unions, the opposing political commitments of their national leaders prevented a permanent liaison.

Perhaps, like the *prima donna,* the amalgamation will have yet another 'final' performance at Council. Essentially the terms for the merger are those appropriate to a loose federation in which DATA retains considerable decision-taking authority, and what amounts to a separate organizational structure at all levels. So long as the constitution of the new federation allows the retention of the draughtsmen's separate identity, it is theoretically feasible that DATA members may return to their former independent status. Furthermore, there seems little evidence to suggest that office committees were enthusiastic in their recruitment of technicians outside the drawing office over the preceding decade.

It is upon the enthusiasm and solidarity of the local office committees that the Association has depended in the past. It seems likely that the strategy and tactics of the newly formed Amalgamated Union of Engineering Workers (the Federation of the A.E.U., the Foundry Workers, Construction Engineers and DATA) will have to rely very much more upon the shared dependency upon external professional negotiators or 'counsellors' to bring together their various skills and statuses. Yet it is precisely the lack of this type of personnel within the other three unions making up the A.U.E.W., particularly the largest organization, the A.E.U., and the type of internal conflict which results from their politically democratic constitutions, that led to a great deal of unintended loss of central control over local bargaining in the engineering industry during the 1950s. From the previous account of DATA's strategy, it will be seen that this lack of disciplined bargaining strategy contrasts with the type of bargaining evolved by the latter.

Yet another problem derives from the notion of DATA becoming the technician section of an industrial union covering the engineering industry. This presupposes that the A.S.T.M.S. does not exist. In practice the latter has been given a limited form of recognition by the Engineering Employers Federation, which allows them to represent supervisors after a ballot of potential members within a particular bargaining unit which gives them a majority. Locally, the A.S.T.M.S. has gained full bargaining rights for both supervisors and technical grades in many large companies in the engineering industry. However, it is precisely in these companies that DATA has represented draughtsmen and the A.E.U. has represented skilled manual workers, many of whom hope to become supervisors, for at least fifty years. Because the Engineering Agreement signed in 1924 constrained DATA's representational rights to the drawing office alone, it is against employers and not against the A.S.T.M.S. itself that the fight for recognition on behalf of other technicians (or 'nons' i.e. non-procedure grades) is directed. The first major dispute was at C. A. Parsons in 1970, when DATA members struck (or were locked out) over a period of four months and ultimately gained the employers' concurrence for exclusive DATA recognition for all technician grades: a status which the union claimed they had gained in an agreement made in 1968.

Shortly after the successful conclusion of this dispute the militant South East Lancashire District Committee of the A.E.F. declared its aversion to working under supervisors who belonged to the A.S.T.M.S. It seems therefore, that the 'leverage' provided by the shop-floor membership of the A.U.E.W. might well become a powerful factor in the future recruiting policy and bargaining strategy of DATA (i.e. the Supervisory and Technical Section of the A.U.E.W.). If this is so, then that part of the A.S.T.M.S. membership which is based in the engineering industry is set at risk. Indeed, if one could regard this alliance as being one between political leaders with malleable resources at their disposal, the future of any potential rivals to A.U.E.W. would seem bleak.

The first factor which stands in the way of the growth of A.U.E.W. across all engineering technician grades is the long-standing reluctance of employers to recognize DATA as being representative not only of technicians of the same status as that of draughtsmen but also of senior staff both within and outside the drawing office. This cut-off point in the representative capacity is made explicit by the National Procedural Agreement[1] for the industry, and is generally adopted by both federated and non-federated employers. As a result DATA has suffered a considerable drain from the most able section of its active membership ever since the agreement has been in existence. The Association has recently adopted the policy of the A.S.T.M.S. in attempting to 'bid up' its services into the ranks of higher-status employees. To quote from a recent exchange between DATA'S General Secretary and a national officer of the U.K. Association of Professional Engineers:

> It is the 'proletarianisation' of graduates which has brought many into DATA in the recent past, adding to the number who have always been members ... Mr. Clarke (U.K.A.P.E.) is inaccurate when he describes our incursion into the £1,800 and £2,000 per year as recent. We have always been there ... In Fords, for instance, we have exclusive negotiating rights for staff up to £3,162 per year and informal rights for those above.[2]

It seems unlikely that many employers will willingly follow the example of Fords. There is some evidence for the belief that many companies have actually preferred to make agreements for their more senior managerial staff with the A.S.T.M.S. even where DATA membership was strong among more junior technical staff. Usually this is simply because they prefer to retain a division in the collective representation of their staff which reflects a division in authority and decision-making within the company hierarchy. There appears to be a growing acceptance among employers of the image projected by Jenkins as the protector of the exposed and helpless middle classes.

Possibly there is also a growing realization that 'actions speak louder than words'—that the A.S.T.M.S. strike record simply does not compare with that of DATA—and an awareness among employees that the A.S.T.M.S. places far more reliance on the technical mastery of the negotiating process than on the assertion of collective sanctions. This may be making 'a virtue

---

[1] Royal Commission on Trade Unions and Employers' Associations, Minutes of Evidence No. 53, ASSET (H.M.S.O., London 1967).
[2] *UKAPE – Unpublished Reply, DATA Journal*, May 1970, p. 19.

of necessity' since the employees recruited by the A.S.T.M.S.—
and those now being sought by DATA—are unlikely to wish to
use the strike action as a normal bargaining weapon. They have
career opportunities and an individual status which they are
unlikely to set at risk in anything but the most unusual circumstances. Unlike draughtsmen, their identity and their actions
cannot be easily subsumed beneath the anonymity of the work-
group. The decision to strike is therefore a much more personal
thing. For these reasons the A.S.T.M.S. members have often accrued a deserved reputation for 'unreliability' among the more militant shop-floor members of the A.U.E.W.

This concern with status among members and potential
members is the second factor which may inhibit the growth of
the A.U.E.W. among engineering technicians. If DATA is now
to attempt to recruit line technicians and more senior grades of
technologists, the internal stresses between the various grades of
its membership, common to all vertically organized 'industrial'
unions, may grow too great for the fragile amalgam of craft
structures making up the A.U.E.W. To the increased heterogeneity in its membership the Technical and Supervisory Section
(formerly DATA) has added the goal of 'unionizing' staff over a
wide hierarchical range of statuses. In order to succeed it must
offer something more than the personal services of a local lay
official. The expertise of professional counsellors who specialize in
the needs of each group of members within the vast body of industrial membership would seem to be a minimum prerequisite for
any organization embarking upon the path of vertical expansion.

Even if it does change in this way, the A.U.E.W. will become
an essentially different organization from that planned by the
leaders of the A.S.T.M.S., who have now quite clearly accepted the
need for a 'specialized union for management and professionally
qualified people'.[1] This does not, of course, preclude the recruitment of foremen, who are seen by the union to be part of
management and who are unlikely to be repelled by this description. Talks with a range of 'protective associations' of professionals were held immediately after the merger between ASSET
and the A.Sc.W. had taken place. The Medical Practitioners
Union, with some 4,500 members, was taken into the Associa-

[1] *The Guardian,* 15 April 1968.

tion, with a separate policy-making and negotiating identity.[1] The Junior Hospital Doctors' Association was offered similar terms, while Mr Jenkins declared that the 'medical profession is ripe for reorganization'.

Nor was this the only profession apparently, for the A.S.T.M.S. went on to hold talks with the Engineers' Guild, the Association of University Teachers, the Association of Teachers in Technical Institutions, and various other protective associations catering for dentists, etc. Clearly the A.S.T.M.S. is 'bidding up' into the range of occupations which A.Sc.W. had abortively attempted to organize in 1917. Its chance of success would now appear much greater since the process of 'bureaucratization',[1] by which professional people are expected to accept the authority and rewards handed down from an anonymous employer, is perceived by many 'professionals' as a threat to their status.

The structure and leadership style adopted by the new union is largely that of ASSET. The amalgamation was achieved by a ballot of members conducted in 1967 in which 11,705 of the A.Sc.W.'s 21,000 book membership participated and 12,500 of ASSET's 50,375.[2] The fact that such a high proportion of the long-term membership of A.Sc.W. voted for the merger demonstrated their anxiety to link up with a large and effective union after their failure to amalgamate with DATA. Their willingness to recognize an effective servicing unit over any personal consideration they might have for institutionalized status symbols was demonstrated at a special Rules Conference held ten months after the amalgamation, which voted to expedite the ultimate merger in inclusive elections by a year, to May 1969 instead of 1970 as suggested by the N.E.C.

For two years members voted six members of the Executive Council from among former members of the A.Sc.W. (Division A), and six from ASSET (Division 1). Joint General Secretary positions were held by Clive Jenkins (ASSET) and John Dutton (A.Sc.W.) until the latter retired in 1970. Now membership is

---

[1] M. Weber, *The Theory of Social and Economic Organization* (New York: Oxford University Press 1947); R. K. Merton, 'Bureaucratic Structure and Personality', *Social Theory and Social Structure* (Free Press, Glencoe 1957), pp. 195–206.

[2] It was made possible by the passing of the 1964 Trades Union Amalgamation Act which enabled such mergers without the necessity of a two-thirds majority of members.

common for most purposes (the exceptions being those eligible for benefits under the old A.Sc.W. schemes). Qualifications for membership include employment as supervisors, scientists and engineers, technicians, managers and executives, students and trainees. In practice these distinctions are not related to separate classes of membership, but, within the union structure, sectional advisory councils are elected by occupational groups, which are allowed a sectional voice in policy-making. The councils are, however, clearly subordinate to the National Executive Council, who appoint Council chairmen and the full-time official who acts as Secretary to the group. The National Executive Council itself consists of eighteen members, twelve elected every two years on a divisional vote, four on a regional vote and two on a national vote.

As well as advancing its claims to organize professional scientists, the A.S.T.M.S. has presented an image of itself as representing business executive interests. In November 1968 the A.S.T.M.S. placed a full-page advertisement in *The Times*. 'A man gets the push because his face doesn't fit. It could only happen at one level in British Industry—at the top.' This was more than a 'gimmick' because at that time the Prices and Incomes Board were investigating executive salaries, and the closure of the A.E.I. factory at Woolwich had made a large number of men in middle management redundant.

It seems possible that the 'bidding up' of the status of their membership appeal has affected recruitment among lesser-status workers more than among the advertised target audience. Thus, while talking to the Association of University Teachers, the A.S.T.M.S. has had rapid and widespread successes in recruiting and negotiating for university technicians. The one-day stoppage by university technicians in June 1969 was the first experience British universities had had of collective action on a national scale, despite the widespread recognition of staff unions which has existed in many universities since the second world war. It was instrumental in bringing about an effective collective bargaining structure within institutions which had previously rejected the basic precepts of institutionalized conflict.

However, more recently the largest gains in membership have been obtained through the peaceful secession of already established company staff associations among clerical workers in the insurance industry. These gains for the A.S.T.M.S. have taken the organization into occupational labour markets in which it has had little negotiating experience and which it has hitherto left to more specialized clerical unions. It was perhaps to be expected that just as an appeal to technologists has attracted technicians, the A.S.T.M.S.' campaign to recruit management should attract clerks. Nevertheless, the union does have a wide range of occupational statuses among the jobs occupied by its members, and, unlike DATA, it has the additional problem of a wide scatter of members over diverse industrial settings. The policy of the A.S.T.M.S. must lead it into competition with other sectorally based unions such as the National Union of Bank Employees whilst in engineering and its other traditional bases its clerical interests have made its relations with the C.A.W.U., G.M.W.U. and A.C.T.S., somewhat ambiguous, to say the least. The problems of the A.S.T.M.S. are those of all organizations which are diversifying at a rapid rate; for the large 'general union' these are to be seen in the need to satisfy the desire for sectoral autonomy and self-expression across the multivarious groups of members, while retaining the centralized control needed for an effective bargaining strategy across a diversity of weak and heterogeneous interests. At the present time the Union boasts a growth rate of 2,000 net a month.[1] It will be difficult to maintain an effective 'counselling' service without a substantial initial investment in overheads. The existing large general unions such as the A.U.E.W. or the Association of Clerical, Technical and Supervisory sector of the T.G.W.U. have such resources. It remains to be seen whether they also have the appeal or, perhaps more important, the expertise to meet the market needs of the growing numbers of technicians.

Conclusion

The history of technicians' unions suggests that this category of workers is strongly status-conscious; but status is seen primarily in labour-market terms. Professional associations that are concerned mainly with non-pecuniary aspects of status do not fully satisfy the strong desires of technicians for higher relative levels of remuneration. On the other hand, appeals to class concepts of trade unionism make little impact. Although the leadership of these unions has tended to be committed

[1] *A.S.T.M.S. Journal*, Issue 3, 1969.

ideologically to left-wing Socialism, and even Communism, the collective bargaining policy actually adopted has reflected the practical requirements of the membership. The ideological wish of the leadership to attack the capitalist system, and the interest of members in securing immediate tangible monetary gains, have come together in a mutually satisfying aggressive policy to secure higher pay and improved conditions of employment.

PART II

PART II

4 The Profile of a Technician

The first part of this book has dealt with problems relating to technicians in general. Chapter 1 discussed changing man-power needs and the supply of technicians; Chapter 2 the prob-lems of identifying and delineating the occupation 'technician'; Chapter 3 was concerned with the growth and development of national institutions catering for technicians. In Part II of the work attention is centred on problems at the level of the enter-prise, and deals with the efforts of technicians to overcome these difficulties and those of management to satisfy the expectations and aspirations of their technician work-force. The basis of the analysis in Part II is provided by the results of a questionnaire survey, together with the interviewing of technicians, and line and personnel managers.

The survey

Before any survey could be undertaken the basic problem to be solved was to decide what collection of jobs constituted the occupational title 'technician'. Chapter 2 has analysed the prob-lems that government agencies and industry have met in attempt-ing to answer this question. In order to arrive at some consistent basis for determining who were the appropriate people to be surveyed the London School of Economics research team talked with companies, unions, professional bodies and government departments. One possible approach considered was that managements willing to co-operate with the study should be asked to compile a list of those employees whom they considered

E

to be 'technicians', but in the light of the problems discussed in Chapter 2 it was decided to reject this approach. Finally a decision was made to use the latest returns made to the appropriate Industrial Training Board by the establishments willing to co-operate. This method does not completely overcome the problem since the I.T.B.s do not have a common definition of the technician. However, it appeared to be the easiest and, in the circumstances, the best method of achieving a reasonably standard criterion for identifying technicians throughout all firms. The next step was to decide upon the sample frame for the conduct of the survey. In the manufacturing sector 85 per cent of technicians are employed in chemicals, metals, engineering, electrical goods, and vehicles, and the concentration of the study was in these areas.[1] The sample chosen was not a random one and the willingness of companies to co-operate was a major determinant of the sample. It was decided to opt for firms in traditional industries, for example iron and steel, but which, nevertheless, were technologically advanced, and those in intermediate industries (i.e. from the point of view of their place in British industrial development), for example the manufacture of motor-car components. A third set of firms was to be chosen from the new and expanding industries, such as aerospace, electronics and oil-refining. With this aim in view fifty firms in steel, aircraft, electronics, vehicles, oil-refining and aerospace were approached for co-operation, and eventually fourteen establishments agreed to give full co-operation to the survey. The product breakdown of these establishments is shown in the table below.

Apart from stratifying the sample on the basis of industry an attempt was made to obtain a representative regional spread of firms within the industries. The aim was to select different parts of the country where the pressure of demand for technician labour was different. The problems of defining the technician labour-market and the difficulties of measuring the state of demand and supply are discussed in Chapter 7. The crude difference between unfilled vacancies and unemployment figures over the period 1954 to 1964 was taken as a guide to the tightness

[1] Manufacturing was chosen because of the better nature of occupational statistics in this sector of the economy. This automatically meant that some significant areas of technician employment were excluded, for example, university, dental, medical and television technicians.

Table 4.1 The sample of companies and products

Industry[1]	Number of Establishments	Products
Metal manufacture	2	Iron and steel
Vehicles	3	Brake linings Engine units Manufacture of aircraft and missiles
Petroleum products	2	Oil refining
Electrical engineering	7	2 Consumer goods 3 Electrical components 2 Capital goods[2]

NOTES
(1) Refers to the Industrial Orders in the Standard Industrial Classification (1968).
(2) Includes computers, radar and navigational aids, radio equipment, and measuring, test and nucleonic instruments.

of the labour market.[1] Co-operation was forthcoming from establishments in relatively tight labour markets in the South-East and North-West, and in the much 'looser' markets of the Northern region, the South-West and Wales. The regional breakdown of establishments is shown below.

South-East	7 establishments
North-West	4 establishments
Northern	1 establishment
South-West	1 establishment
Wales	1 establishment
TOTAL	14

The sample firms were chosen not only by industry and region, but also with a view to differences in production systems. Woodward,[2] on the basis of a study of industrial organization in South Essex, distinguished three major production systems—unit and small-batch, large batch and mass production, and process—with a series of subdivisions within each so that

[1] This refers to the Department's Standard Regions for statistical purposes. There are ten such regions. See the *Ministry of Labour Gazette*, January 1966, p. 20.
[2] See Joan Woodward, *Industrial Organisation: Theory and Practice* (Oxford University Press, 1965).

there were eleven systems altogether. The reason for considering technology factors was that these might be a variable explaining the behaviour and attitudes of technicians. For example, it seemed conceivable that where a large number of technicians were employed in a research and development function in the manufacturing process and this was a critical activity, a higher status might be accorded to such groups by the management. The influence of the type of organization on technician behaviour is discussed in Chapter 5.

The sample frame (see Table 4.2) was finalized and a pilot study was conducted in an iron and steel firm in the summer of 1967. The open-response interviewing of technicians, union representatives, line and personnel managers used in the pilot was conducted during the summer and autumn of 1968. Interviews at the place of work were held with key persons in the unions as well as with individual technicians. In order to execute these interviews, together with those of managers, a period of between one and two weeks was spent at each establishment, including some time for acclimatization and observation. The overall response rate to the questionnaire was 60 per cent, but replies from firms in London and the South-East in electrical engineering were below this average.

The results of the survey

From the survey it was possible to distinguish ten groups of technicians, of which there were four main groups—draughtsmen, laboratory staffs, planning and production engineers and quality-control technicians—which, despite having marked differences of job task, had some very similar characteristics. It is in the light of these differences and similarities that this profile of the technician was built up.

Types of jobs performed

Although the research team did not make consideration of the job specifications of technicians a major subject of investigation, it was necessary to make an analysis of the type of jobs performed. Technicians, unlike skilled and unskilled manual workers in manufacturing (except for those on maintenance), are not doing jobs involved in direct production; their tasks are confined to servicing the establishment's processes of production. Although technologists are also not involved in direct production they carry in the performance of their jobs, in the

Table 4.2   The sample – industrial and regional breakdown

Industrial order (1)	Region	Number in sample	Number of questionnaires mailed	Number of questionnaires returned	Percentage response (2)
Petroleum products	South-East	80	80	48	60
	North-West	260	260	172	66
Metal manufacture	Northern	80	80	53	66
	Wales	347	347	201	58
Electrical engineering	South-East	533	533	286	54
	North-West	177	177	117	66
	South-West	124	124	83	67
Vehicles	North-West	50	50	32	64
	South-East	294	294	175	60
Total		1,945	1,945	1,167	60 (3)

NOTES
(1) Industrial Orders are based on the Standard Industrial Classification (1968)
(2) Figures in this Table and all Tables following are rounded to the nearest whole number
(3) This figure is the average overall response rate rounded to the nearest whole number

majority of cases, a greater degree of responsibility. In only a few cases were technicians actually carrying out a job without being responsible to somebody else. They worked in small groups, and although often required to exercise technical judgement this was done on the basis of understanding the general principles, reasons and purposes of their work, rather than on that of skills developed from experience. Technicians were also different from technologists and skilled manual workers in that they could not be identified on the basis of a single national qualification. Engineers and technologists have been identified in the reports of the Committee on Manpower Resources for Science and Technology on the basis of an educational qualification of degree or equivalent level, or some well-established professional qualifications. Skilled manual workers are identified by the common characteristic of a craft apprenticeship. For technicians, however, the problem is complex in that some have degrees, some O.N.C., others City and Guilds Certificates, and some no nationally recognized technical qualifications at all. It is for this reason that attempts to define technicians in the past have taken a functional approach.

The survey enabled four main groups of technicians to be identified on the basis of task performed. These four groups accounted for 80 per cent of the sample and it is upon these respondents that this analysis is concentrated. Draughtsmen were engaged on design, the ironing out of faults in design before production, development work, and estimating. Laboratory staffs were involved in routine analysis and routine testing, using laid-down sample schedules following known and proved techniques. Some were also involved in research and development, either in roles supporting research officers or in jobs requiring a smallish degree of responsibility. Planning and production engineer technicians performed a number of tasks. Some were concerned with work study, some with method study, and others with production planning, but the work was generally routine, being carried out within a basic set of rules. Quality-control technicians were involved in testing and inspection; inspectors were concerned with checking output from production departments before sending the product to the testers. For example, in electronics firms they would inspect wiring of circuits. Some inspectors, however, performed a maintenance function in that they tested out new materials and checked plant before use in production runs. The testers were engaged in testing the finished product before release to the customers, and one of their major functions was to find faults.

Despite similarities in the functions of technicians in the production process, the differences in the tasks performed appear to have prevented the emergence of an occupational identity as 'technicians'. In reply to a question 'If someone you had never met before asked you your occupation, what would you say?' only 2 per cent answered 'technician'. Draughtsmen differed from other groups of technicians in that they had a strong sense of belonging to the occupation 'draughtsman'. In reply to the same question 65 per cent answered 'draughtsman'. This is not altogether surprising, since draughtsmen tend to be employed in groups in the drawing office, isolated from other groups in the plant, and this has naturally led to a feeling of identity. The trend is perhaps also reinforced by the fact that draughtsmen are the only group of technicians to have a national trade union catering for them. Although laboratory technicians generally work together in a laboratory, as do draughtsmen in a drawing office, the group does not appear to have developed a high degree of self-identification. Many laboratory technicians described themselves as chemists, and in iron and steel plants, as metallurgists. In answer to the question 'How well do you think the term "engineer" describes you in your work?' laboratory technicians, unlike draughtsmen and planning and production engineers, considered the term engineer a bad description. Thirty-nine per cent of laboratory staffs believed the term 'technician' described them very well or well, as opposed to 18 per cent who thought it a bad description. Quality-control technicians showed a slightly greater readiness to accept the term 'technician'; some 48 per cent agreed that it was a good description of their work role.

Personal characteristics

Technicians are young, predominantly male, and married: 71 per cent of draughtsmen, 75 per cent of laboratory technicians, 62 per cent of planning and production engineers and 68 per cent of quality-control technicians were under forty years of age. In this respect technicians are little different from technologists and skilled manual workers.

Education and training

The most common type of education received by technicians, apart from draughtsmen, was grammar-school, and amongst laboratory staffs 56 per cent attended such schools. Secondary-modern schooling was most frequent amongst draughtsmen, but nearly a quarter had attended secondary-technical schools. The type of schooling received by technicians separated them from manual workers, the majority of whom had attended secondary-modern schools, and many of whom had received no secondary schooling. Technologists were similar in that the grammar school was the most common form of schooling, although the percentage attending such schools was probably higher than amongst technicians. For all four groups G.C.E. O level is the most common school attainment of technicians. The proportion of those reaching advanced level G.C.E. was highest amongst laboratory staffs and planning and production engineering technicians.

A number of studies[1] of technicians have shown the level of formal technical qualifications for their jobs to be low, but the questionnaire survey conflicts with these findings: 72 per cent of draughtsmen possessed a formal qualification, as did 56 per cent of laboratory staff; 72 per cent of planning and production engineers, and 57 per cent of quality-control technicians. The London School of Economics study probably illustrates a catching-up process in the sense that when the previous studies were undertaken technicians were probably in the process of gaining technical qualifications. For all four groups in this study Ordinary National Certificate and Higher National Certificate were the formal qualifications most frequently found, but the value of replies was limited in that a large number of technicians did not answer the question.

Technicians appear to have had a wide range of formal training involving on-the-job and off-the-job experience, and provide examples of both general and specific training. By general training is meant the acquisition of skills that are useful to more than just the firm providing the training, whilst specific training refers to the accumulation of skills that are of application only in the plant or company providing the training.[2] Types of general training received included craft apprenticeships, craft apprenticeship plus office experience, drawing-office apprenticeship, and training in the forces. Specific training included company technician-staff training schemes. With the growth and development of Industrial Training Boards a great deal of on-the-job training for technicians is being recommended by the Boards. In the engineering industry the I.T.B. has designed a joint first-year course of training for technicians and craft

---

[1] S. Cotsgrove, *Technical Education and Social Change* (Allen and Unwin, London 1968). Committee on Manpower Resources for Science and Technology, *Report on the 1965 Triennial Manpower Survey of Engineers, Technologists, Scientists and Technical Supporting Staff*, Cmnd 3103 (H.M.S.O., London 1966).

[2] Alternatively, types of training can be defined in relation to the effects on the marginal productivity of labour. General training will raise the marginal productivity of all workers, while specific training would raise the marginal productivity of the workers in the firm providing the training, and not in other firms. This classification is essentially a conceptual one and is difficult to apply in practice. See G. S. Becker, *Human Capital* (National Bureau of Economic Research, Columbia 1964).

apprentices which covers the major part of on-the-job training. In subsequent years the technicians undertake particular types of training while the apprentices choose two of a number of set modules.

Source of recruitment

The majority of technicians in the survey had been on the technical staff of their present company for less than ten years. Part of this is explained by the relatively young age of the majority of the sample and partly by the fact that, apart from laboratory technicians, most technicians in the survey had joined the technical staff of their present company from that of another. This implies labour mobility among technicians, and an analysis of the extent of this is undertaken in Chapter 7. There are two schools of thought concerning the recruitment of technicians.[1] The 'old school' argues that recruitment should be from the best craft apprentices. A firm takes on a number of apprentices, then at the end of their apprenticeship, and sometimes during it, the best ones are encouraged to join the technical staff and the others to remain on the manual staff. The 'new school' believes that the recruitment of technicians should be direct into company technician-training schemes. In other words the firm should employ school-leavers on technician courses from the beginning and not bother with apprenticeship training. The survey showed that for the technicians in this study the majority had been recruited according to the 'old school'. Laboratory technicians appeared to be recruited direct from school, and upgrading from the shop-floor was rare: 37 per cent had joined their present company from school, and this was greater than the number of laboratory technicians who had come from other firms.

It would appear that technicians are recruited from school into craft apprenticeships, and that the brighter ones are upgraded to the technical staff and then encouraged to obtain formal technical qualifications, like the Higher National Certificate. This recruitment pattern differs from that of the technologist and the manual worker. Although many technologists have arrived at their present positions through promotion, as a result of experience and part-time study most have been recruited from universities, or colleges of advanced technology, and then given a

[1] See Chapter 1.

short period of on-the-job training before assuming full responsibility for tasks. The study found this to be a source of discontent amongst technicians, which was revealed in comments like: 'University types come in and we have to tell them the job and then they are put in charge. Our experience counts for nothing.' The skilled manual worker is similar to the technician in that he is recruited straight from leaving secondary school, but is different in that after completing his apprenticeship little extra training and little encouragement is given to obtain higher qualifications. Off-the-job training and further education is usually only required when craftsmen wish to keep up to date with new developments in their crafts.

Family background

In all but one occupational group in the selected population, technicians from manual working class homes constituted 50 per cent or just under of respondents. The exception was the quality control group in which only 34 per cent were sons of manual workers: these latter line technicians along with planning and production engineers were rather more likely to come from the homes of clerical workers. These latter groups, who it will be remembered were occupying a more 'managerial' position than that of most other technicians, were also more likely to claim parental origins among the professional and executive classes.

Membership of occupational organizations

Technicians are the most unionized of white-collar groups in the manufacturing sector, with a density of 30 per cent, compared with an overall white-collar average of 12 per cent. The survey returns showed marked variations in unionization by occupation; 56 per cent of laboratory technicians were organized, 43 per cent of draughtsmen, 37 per cent of quality-control technicians, and 22 per cent of planning and production engineers. The percentage of respondents who had never joined a union varied widely between groups, but nearly a third of draughtsmen and a quarter of laboratory technicians had been members of a union for technical staffs for more than ten years.

Membership of a professional institute or association for

Table 4.3 Occupations of technicians' fathers

Occupation	Skilled Manual	Un-skilled and semi-skilled Manual	Routine Manual worker	Super-visory and inspec-tion	Profes-sional and execu-tive	Own business	No. answer
Draughtsmen	22	26	14	6	10	6	17
Laboratory staff	27	23	11	12	9	7	11
Planning and production engineers	22	28	21	4	16	4	5
Quality control technicians	17	17	19	10	13	7	19

NOTE
(1) Figures are in percentages. SOURCE: Questionnaire Survey.

technical staff was low among the technicians. This was not surprising in that few in the sample had the minimum qualifications to gain entry to such a body. Planning and production engineering technicians were an exceptional case in that 30 per cent of them were members of some professional body. Over 80 per cent of draughtsmen and laboratory technicians were not in membership, and the figure for quality-control technicians was over 90 per cent. Technologists, scientists and engineers, unlike technicians, form the bulk of membership of such institutions, while trade-union membership is much smaller than among technicians. The much lower degree of unionization amongst technicians than amongst skilled manual workers, where in some trades membership is virtually 100 per cent, helps to distinguish technicians from these occupational groups. It has been suggested that technicians have probably shied away from unions because membership of such bodies has seemed incompatible with being on the company staff, and hence part of the management team. However, unionization of technicians has been increasing at an exponential rate, and management is finding itself faced with collective action from a group of workers previously thought to be more concerned with solving their problems on an individual basis. How the managements in the establishments in this study have reacted to this problem is dealt with in a later chapter in this book.

Attitudes at the work-place

The survey showed technicians as receiving satisfaction from a number of features at the work-place. All four of the occupational groups expressed satisfaction with the proper utilization of their technical skills, with the interest of their work and with the people they worked with, while only about 20 per cent thought identification with the aims of the company brought them dissatisfaction with their job. Despite a high degree of satisfaction in their work, all four groups appeared to agree that they were not part of the management team. The feeling was strongest amongst the planning and production engineers, of whom 42 per cent replied that they either strongly or very strongly disagreed with the statement that they were part of the management team. As regards the competence of the management at the plants in the survey, technicians' attitudes varied according to the area of management concerned. All four occupational groups thought that their managers had a good grasp of the technical problems with which they were faced, but a poor understanding of human relations problems. Draughtsmen and planning and production engineers considered that management took considerable interest in their problems as technicians, but laboratory technicians and quality-control technicians held the opposite view. These attitudes would appear to fit into the popular view of British management, namely its neglect of human relations and industrial relations and its predominantly production orientation. Line management has shown little concern with man-management problems, but some of the firms participating in this study were showing distinct signs of change. One company had already started to try and change the attitude of line management. Chapter 9 gives a more detailed analysis of managerial attempts to overcome these problems.

Technicians had strong views on the status accorded to them by management. Only one-third of all groups considered their status to be satisfactory; the feeling that it was definitely not high enough was strong amongst draughtsmen (52 per cent), laboratory staffs (56 per cent) and quality-control technicians (57 per cent), but less than 40 per cent of planning and production engineers held similar views. The state of morale amongst technicians appeared to be average or low. For all occupational groups there was a low number of returns describing morale as high, and the most common assessment of morale was that it was about average. For laboratory technicians, however, a higher percentage said morale was low or very low, rather than average. The reason for this low morale is complex, and the chapters that follow provide some explanation.

Aspirations of technicians

The survey revealed technicians to be an ambitious group. Over two-thirds of all replies from the major occupational groups showed an expectation of promotion to another job, and over four-fifths a positive ambition for further promotion. Most of the four groups of technicians appeared not to be satisfied with what they had already achieved, and the desire to become a manager one day is a not uncommon hope of technicians. One of the themes to emerge from this study is that it is the failure of management to satisfy these aspirations and expectations of technicians that explains some of the behaviour patterns of this occupational group. The aspirations and expectations stem from the fact that technicians accept a common perception of the importance of possessing educational qualifications, namely that they are a passport to a good and secure job. Here a 'good job' means one with prospects of career advancement. Many technicians have acquired H.N.C. and O.N.C. through part-time study at the cost of forgoing leisure time and the opportunity to supplement income in the evenings, for example by serving behind the bar in a public house. Education has been seen as an 'investment good' from which over a period of time a rate of return at least equal to the cost of acquiring the 'good' is expected.[1] This attitude towards education is fostered from schooldays: teachers in grammar schools impress upon pupils the need to pass examinations, to attend an institution for higher education and then obtain an appropriate post in industry, etc., and in secondary-modern schools the emphasis is placed on obtaining a craft apprenticeship, the 'Get yourself a trade, lad' attitude.

---

[1] Those not familiar with the 'rate-of-return' approach to education are referred to the following:

G. S. Becker, *Human Capital* (National Bureau of Economic Research, Columbia 1964).

M. Blaug, *The Rate of Return from Education* (The Manchester School, 1965).

or M. Blaug (ed.), *The Economics of Education 1* (Penguin Modern Economics Series, 1968).

Technicians, then, have developed expectations about future income-streams in relation to those who have fewer units of human capital, and about the life-styles they should adopt. When these do not materialize, frustration develops.

It is not easy to pinpoint the reasons for this non-materialization of expectations, and the study has not involved itself deeply in this problem; instead it has concentrated on the question of how technicians and management have reacted to the situation. The answer probably lies in a greater readiness to substitute technicians for technologists, and skilled manual workers for technicians but to make the upward promotion of technicians more difficult. The new situation stems from the application of technological change, which so alters the skill-mix of existing jobs that new jobs are created and present ones destroyed or severely modified. Unfortunately there are immense problems of measurement in using the concept of elasticity of substitution between factors of production. Impressions gained from the research strongly suggest that the degree of substitution of technicians for technologists has lessened, but that the substitutability of skilled manual workers for technicians has become easier. Hence it is increasingly difficult for a person to join a company as an apprentice, progress on to the technical staff, and then through part-time study and on-the-job experience graduate up to the grade of technologist and the junior levels of management. The technician is under pressure from two sides. Entry into the technologist grade is becoming blocked as employers demand a degree or equivalent qualification as a minimum hiring standard. The skilled manual workers are extending into technician jobs through the enlargement of their tasks. In one firm participating in the study the laboratory technicians had lost over 100 routine jobs to plant operators, while in another the work content of fitters had been extended into technician areas, as a means of increasing productivity, in order to give the group a pay rise that could be justified under the existing criteria for pay increases under the Productivity, Prices and Incomes Policy.

The frustration of the technicians was illustrated in their attitudes to promotion opportunities in their companies. All four groups disagreed or strongly disagreed that there were sufficient promotion outlets in their company, with only about 20 per cent considering opportunities for career advancement sufficient. Managements in the plants, however, appeared convinced that promotion opportunities were adequate. Chapter 6 deals with the levels of salary amongst technicians in detail, but it is sufficient to say, at this stage, that technical supporting staffs were dissatisfied with their salary level, although about one-third were indifferent. Feeling was strong on the relative levels of pay, where all groups emphatically replied that the pay of technical staff, compared to that of shop-floor workers, was insufficient. Technicians apparently perceive their additional units of human capital over manual workers as being insufficiently rewarded in relation to the cost paid to acquire these extra units.

Conclusion

The questionnaire survey revealed a number of common characteristics of technicians, even though there were marked differences in their job tasks.

(1) Technicians are young. Three-quarters of draughtsmen and laboratory technicians were under 40. The 'older' technicians were to be found among planning and production engineers, of whom 38 per cent were over 40.

(2) Technicians tended to be mainly the sons of manual workers.

(3) The most common type of schooling amongst technicians is grammar-school (laboratory staffs 56 per cent, draughtsmen 29 per cent, planning and production engineers 43 per cent and quality-control technicians 34 per cent). The majority of technicians possess some formal technical qualification for their present job. Higher National Certificate and Ordinary National Certificate are the most frequently found formal qualifications.

(4) Technicians have been on the technical staff of their present firm for less than ten years, and a large percentage have moved to their existing job from another firm. The major source of recruitment of technicians appears to be from the ranks of craft apprentices, but for laboratory technicians direct entry from school into the lab is the norm.

(5) Technicians are union-organized to a greater extent than any other white-collar employees in manufacturing, with a density of 30 per cent. Three of the four occupational groups in this survey had a higher density. Laboratory staffs were 56 per cent unionized, draughtsmen 43 per cent, and quality

controllers 37 per cent. Membership of professional bodies or associations for technicians was limited.

(6) Technicians are satisfied that generally there is a proper utilization of their technical skills; they are interested in their work, and they identify with the aims of the company and the people with whom they work. However, they feel that the status accorded to them by management is too low, and in spite of their satisfaction in the jobs they are doing, the morale of the technicians is also low. Technicians, as a group, appeared to have little feeling of being part of the management team in their companies, and have a poor opinion of management's grasp of human-relations problems. They do, however, have a high opinion of management's grasp of technical problems.

(7) Technicians are an aspiring occupational group. There is a strong desire for promotion to another job in the same field, a burning ambition to achieve promotion, and little satisfaction with what has already been achieved in career advancement. The high aspiration is also seen in their strong hopes of joining managerial ranks at some time in the future. In this survey it appeared that these aspirations were not being satisfied, and there was strong feeling among technicians concerning the promotion opportunities available: 65 per cent of draughtsmen, 68 per cent of laboratory staffs, 54 per cent of planning and production engineers and 61 per cent of quality-control technicians disagreed or strongly disagreed that there were sufficient promotion opportunities in their companies. The other area of strong dissatisfaction among technicians was the differential in wages between themselves and those on the shop floor.

Technicians are young and aspiring, and it is the failure of management to provide for these expectations and aspirations that helps to explain the worsening industrial-relations situation between management and technicians. The frustration of the technicians probably stems from the problems of adjusting to change, whereby the skills required for an occupation have altered. This study is concerned with ways in which technicians have attempted to overcome these problems of blocked career-advancement and declining relative salaries, and the response of management to these actions on the part of technicians.

5 Orientation to Work and the Work Situation

Within the occupational categories chosen for analysis the differences between the roles played by technicians in the work-place were considerable. The reasons for this were multifold. Organizational, technological or purely economic factors determined the physical division of labour within the work-place, but in addition, a more subjective element was added by the readiness of management to give certain categories of technicians different titles as an indication of difference in status. While it has been found expedient to ignore such titular details in our statistical aggregations, they were often a most important causal component in determining the attitudes of technicians briefly described in the previous chapter and analysed in greater depth in this one.

It is proposed to examine the nature and degree of the satisfactions derived by technicians within their work situation and to relate these to their expressed needs. In this chapter the importance of the organizational setting, the technology and the product market in which the employing firm is operating will be emphasized, but not to the exclusion of the propensities which were brought to the work-place by the employee himself. It will be suggested that while the immediate work-environment is important in determining the attitudes of those in the sample, it is erroneous to isolate this from the history and career expectations of the individual technician, and from the groups, internal and external to the work organizations, which influence the expectations of technicians. It is also important to bear in mind the fundamental influence of the labour-market position of the technician in determining his underlying fears and insecurities.

The technician and his job

A clear distinction has often been discovered between the orientation to work displayed by white-collar employees and that of manual workers. In the findings of studies conducted by sociologists and psychologists, occupational differences in the areas of work satisfaction have tended to lead to and to be synonymous with the conclusion that the needs brought into the work situation were also qualitatively different. Thus one American sociologist was led to explain that, 'To some extent, these findings on occupational differences reflect not only differences in the objective conditions of work for people in various jobs, but also occupational differences in the norms with respect to work attitudes.'[1] The white-collar employee has been seen to be a more work-oriented person[2] than the manual employee, and a person for whom the work role was central to his own self-identity. Such a person derives his greatest satisfactions from the fulfilment of his task-centred needs. He is rewarded by the intrinsic value of a job well done, by interesting work, and by the proper utilization of his skills, as against the purely extrinsic factors such as salary and fringe benefits. He may or may not derive satisfaction from his physical working conditions and from his companions, but unlike the shop-floor employee, he attaches less importance to these factors than to those of achievement and self-fulfilment in his work role.

If the degree of satisfaction recorded in Table 5.1 reflects the current preferences of the technicians replying to the question, it seems to suggest that technicians are highly job-centred individuals. Their interest in their work appears to provide a major source of psychic rewards, a source which contrasts quite markedly with the degree of job satisfaction derived from more extrinsic aspects of their position in the firm such as salary and fringe benefits. On the other hand the amount of importance

[1] Robert Blauner, 'Work Satisfaction and Industrial Trends in Modern Society', in Walter Galenson and S. M. Lipset (eds.), *Labor and Trade Unionism*, (Wiley, New York 1960), p. 343.
[2] Robert Dubin, *The World of Work* (Prentice Hall, Englewood Cliffs, 1958), pp. 254–58. See also 'Industrial Workers' Worlds: A Study of the "Central Life Interests" of Industrial Workers', *Social Problems*, Vol. 3 (January 1956), p. 131.

attached to the support and friendship of their colleagues also appears extraordinarily high in this and in other answers to questions asked in this study, particularly among quality controllers. It seems that work-group interaction is a much more immediate and more important means of integration with the organization than is suggested by an identity with the aim of the company itself. The majority of technicians were at best indifferent to the kind of 'moral involvement'[1] implied by such an identity. There were, however, important differences in the degrees of satisfaction, which were not revealed in the 'scores' shown in Table 5.1. Among the young laboratory staff only 28 per cent derived any satisfaction at all from this aspect of their job, and barely more among draughtsmen (32 per cent). But significantly more were found among quality controllers (39 per cent), and over 40 per cent of planning and production engineers found satisfaction in this kind of identification with the company.

Table 5.1 Which of the following factors bring you satisfaction in your present job?

Factor	Draughtsmen	MV	Laboratory staff	MV	Quality controllers	MV	Planning and production engineers	MV
1. The people you work with	73	0.70	73	0.73	76	0.85	73	0.82
2. The proper utilization of your technical skills	72	1.35	67	1.55	71	1.51	69	1.31
3. Your salary	57	1.20	51	1.15	54	1.37	57	1.28
4. Your fringe benefits	57	1.33	56	1.35	65	1.79	60	1.50
5. Your interest in your work	79	0.90	79	0.94	82	0.72	81	0.95
6. Your identity with the aims of your company	63	1.38	61	1.31	66	1.87	66	0.99

MV = mean variance   SOURCE: Questionnaire Survey

The scores in Table 5.1 are computed from a weighting given to a five-point scale of satisfaction from 'Much satisfaction' to 'Much dissatisfaction' under each factor heading. They are therefore an expression of the relative degree of satisfaction or dissatisfaction, and *not* a proportion of the sample.

The perceived needs of the technician were tested in a number

[1] The term 'moral involvement' is used here as a concomitant to 'normative authority' as in A. Etzioni, *A Comparative Analysis of Complex Organisations* (Free Press of Glencoe, New York, 1961). It contrasts with a calculative involvement which indicates that the person is only involved to the extent of doing a 'fair day's work for a fair day's pay' and regards the authority of the company as being limited and 'utilitarian'.

of ways, among which the forced-choice question set out below
was one. It became apparent that for the lowly paid groups
higher financial rewards were given a paramountcy over potential
intrinsic returns to be gained through their single allowed choice.
Yet even among the relatively better-paid planning and production engineers, as well as among draughtsmen, higher salaries
were given first priority by significantly large proportions of this
sample.

**Table 5.2   What changes would you most like management to make
in order to give technical staff more satisfaction in their job?**

Changes	Draughts-men	Labora-tory staff	Quality control-lers	Planning and production engineers
More opportunities for promotion	19	18	15	16
Better use of technical skills	28	16	21	23
Higher status in the company	17	17	19	19
Higher salaries	22	34	32	24
Better fringe benefits	1	3	1	3
Free-choice answers { Combination of these factors	9	9	7	10
Better working conditions	1	1	1	0
More technical education	1	0	0	0
Don't know	2	2	4	5
Total	100	100	100	100

NOTE: figures are in percentages   SOURCE: Questionnaire Survey

This fact seems to detract, not only from the notion that
white-collar workers are relatively less aware of extrinsic rewards
than are manual employees, but also from the possibility of
there being some absolute level in the satisfactions to be derived
from meeting such 'lower order' financial needs. One school of
industrial psychologists has put forward the notion of a hierarchy of human needs brought into the work-place by every
employee, ranging from the purely physiological, through the
universal need for social recognition and identification, to the
need for self-esteem and recognized achievement in a chosen
occupation. It has been further suggested that the basic needs
of the employee, as represented in salary and working conditions,
in interpersonal relations and in supervisory styles, or in the

company policy and administration, can only be satisfied in some partial sense. The truly satisfied person is one who is motivated by achievement in the task, by recognition, and by personal advancement and increased responsibility.[1]

The proponents of this theory and the policy implications that flow from it generally begin by assuming an affluent and fully employed work force which has gone some way towards satisfying the individual need for security, increased salary, and other 'lower order' or 'hygiene' conditions. The list of attributes regarded as 'motivating' are curiously specific to those that might be regarded with approval in an individualized Western society. The manager is therefore seen as being able to concentrate his strategy on developing the task content of the jobs within his organization. It is only through enriching the job with increased variety and, more importantly, responsibility, that he will acquire well-motivated employees.

This approach has had some impact on British management in large corporations, and indeed appeared as part of the formal policy of at least one large company co-operating in this survey. Its emphasis is on individual rather than collective advancement, and is therefore particularly appropriate to an aspiring and some-what heterogeneous category of employees lacking a clear sense of identity, such as technicians. The technicians in this study were extremely ambitious, the lowest level of aspiration being among quality controllers, where only 74 per cent desired promotion. Furthermore the vast majority of technicians preferred methods of payment which included rewards for individual merit (see next chapter).

Yet taken at face value it appeared that management were singularly failing in their attempts to meet the technicians' expectations of their work role and status. Morale was not high, particularly among laboratory staff and draughtsmen. This was associated with the belief that promotion for technicians in their employing firm was insufficient, a view which was again most strongly held in the drawing office and laboratory. When asked: 'For what reasons do technical staff receive promotion in this company?' laboratory assistants gave precedence to technical

---

[1] For the most sophisticated approach to the two-dimensionality of human needs see F. Herzberg, B. Mausner and Barbara Synderman, *The Motivation to Work* (Wiley, New York 1959). For a different treatment see C. Argyris, *Personality and Organization* (Harper, New York 1957).

qualification, planning and production engineers to 'experience' and draughtsmen and quality controllers to 'being at the right place at the right time'. In themselves these views express a realistic view of the major constraints on promotion within the sample firms. For laboratory technicians these included the influx of well-qualified technologists which then formed a barrier above them. For draughtsmen the limited number of senior design positions made promotion within the drawing office an uncertain prospect, and for the quality controller also, promotion into management seemed remote. On the other hand the planning and production engineer had often already gained promotion from the shop floor or drawing office on the basis of his past experience, and many regarded this achievement as a legitimate basis for future promotion.

There were differences in the degree to which technicians accepted the notion of blocked promotion opportunities, and it appeared that most saw these as capable of being remedied by management. However, it was clear that their dissatisfaction was more closely and immediately related to the perceived under-utilization of their skills and to the level of their earnings. Some of the reasons and attempted remedies for the first of these problems are examined later in this chapter. For the present it is worth noting the close relationship between the relative level of earnings and the technicians' view of their status (see Table 5.3).

This relationship reveals a weakness in any theory or policy that begins from a belief in some static notion of a sufficiency of extrinsic rewards needed to satisfy the basic needs of the individual. To begin with, the individual does not carry out his work role in a vacuum; he is constantly judging his task, role and rewards against a number of reference points, and is more or less satisfied in relative terms only. Secondly, the work-group does more than offer rewards of ascribed status and approval. People working in similar conditions will come to be recognized as equals or peers and will normally be regarded as colleagues. Such individuals may develop an identity arising out of their common conditions of work and shared expectations and senti-ments. They may or may not constitute a work-group depending on the nature of the task and geographical proximity. Draughts-men, for example, had developed an extremely high sense of normative identity within the place of work, and an intense loyalty to their colleagues, and had come to see their personal standards, their level of achievement and their own definition of ego-fulfilment in group terms as well as in terms of purely individual ambition. These standards became institutionalized, and eventually an interest-group such as a draughtsmen's association may be formed to protect such standards against violation by an employer or other 'outsider'.

Table 5.3 In your opinion how can management best demonstrate to technical staff that they are considered important within the company?

Occupational group	By emphasiz-ing the value of qualifi-cations	By providing scope for promotion	By main-taining good pay differ-entials	By keeping in close contact with the work and problems of tech-nical staff	Open-ended answer— yielding a combina-tion of these factors	Total
Draughtsmen	4	22	40	30	3	100
Laboratory staff	4	25	48	20	3	100
Quality controllers	6	12	30	44	2	100
Planning and production engineers	5	30	16	44	1	100

NOTE:
Figures are in percentages.   SOURCE: Questionnaire Survey.

A colleague group does not of course have to become the sole source of behavioural norms for its members. It may be fairly loosely structured and consist of people who simply have more in common with each other in given roles than with all others playing different parts in the situation. Many groups of technicians, such as quality controllers and planning and production engineers, may be constrained by their work role and the technology with which they work from taking part in a great deal of social interaction with those regarded as colleagues within the work-place. Yet they may have a real sense of identity as a group which becomes manifest under stress, such as when any one of the loosely defined expectations or aspirations, developed within their work role, has become accepted as legitimate by their group, and is challenged by an outsider.

In the ambiguous situation which technicians hold in the local status hierarchy of the work-place it has been suggested (see Chapter 2) that they may relate either to management or to the shop-floor employee for guidance as to what is appropriate within a situation. From either one of the larger and socially more significant groups the 'men in the middle' may gain a fairly well-formed code of behaviour as a guide to their collective beliefs and actions. If, however, the individual member perceives himself as having a good chance of promotion out of the group into management, it is unlikely that he will choose shop-floor employees as his normative reference group. This is particularly so if he has already experienced promotion into his present position and if management have themselves once used the promotion route that he is now treading.

Table 5.4   **Would you like to be a manager some day?**

Occupational Group	No answer	Yes, very much so	Yes, would quite like to	Not really	No	No, would very much dislike to	Don't know	Total
Draughtsmen	1	28	29	20	17	4	1	100
Laboratory staff	2	29	32	16	12	7	3	100
Quality control	4	24	26	15	23	6	2	100
Planning and production engineers	6	33	34	8	11	6	2	100

NOTE: Figures are in percentages.   SOURCE: Questionnaire Survey

It is clear that despite the apparent obstacles many technicians saw their career in these terms, and in Table 5.5 it is shown that for whatever reason, management as a normative reference group was very much more important to the technicians than was the shop-floor employee. Many technicians will be motivated no matter how indirectly to model their values and attitudes upon those of management as part of this preparatory process. Others may simply wish to be identified with the higher group, although no objective hope of promotion exists for them any longer. Whether the process is regarded as preparation for management or simply a choice based on desires for a higher social status, technicians are overwhelmingly inclined to identify with management.

Table 5.5 Would you say that, in general, technical staff have an outlook more in common with management or with shop-floor workers?

Occupational Group	Definitely more in common with management	Somewhat more in common with management	Neither	Somewhat more in common with shop floor workers	Definitely more in common with shop floor workers	It varies	Don't know	Total
Draughtsmen	13	43	15	16	2	10	2	100
Laboratory staff	10	41	17	16	6	11	1	100
Quality controllers	11	42	19	13	3	12	1	100
Planning and production engineers	20	41	15	8	1	11	4	100

NOTE: Figures are in percentages.　　　　SOURCE: Questionnaire Survey

Whether attempting to maintain their perceived status or to improve it, the technician's position may be measured on one hand by the degree of 'acceptance', or by the treatment that management metes out, and on the other by reference to the degree of social distance technicians put between themselves and shop-floor workers. They are extremely sensitive to any indication of their differential position in the firm, for example in levels of pay, working conditions, supervisory treatment, etc. Technicians constantly compare their earnings and conditions with those of the shop-floor workers, whose relative position is a matter of continuing concern. In terms of personal achievement technicians may refer to their employer and relate their rewards to those achieved by management, but in terms of the extrinsic rewards, so important to group status, the principal comparative reference is generally the earnings of the shop-floor employees.

These concerns are illustrated by the data in Table 5.6, and they are more conclusively supported by the case histories in technological change at the end of this chapter. The attitude of technicians may be rationally explained if the skills of the technicians are seen as a 'capital resource' upon which each technician is attempting to maximize returns over his lifetime. This may of course include the desire not only to maintain differential earnings, but to increase them, and to be considered as occupying a managerial position. Yet without taking account of the need for status satisfaction it is difficult to explain some of the seemingly irrational acts leading to open conflict in the work situation and the recent changes in form of technician behaviour. This is particularly shown in the next chapter, which demonstrates that many of the technicians who were so definite in their views on the injustice of the erosion of their pay differentials were no better qualified in terms of formal training than the shop-floor workers.

Table 5.6 Do you think that the pay of technical staff is insufficient compared with that of shop-floor workers?

Occupational Group	No answer	Definitely yes	Yes	Not sure	No	Definitely no	Don't know	Total
Draughtsmen	I	55	32	4	4	I	2	100
Laboratory staff	2	62	22	5	7	2	0	100
Quality controllers	2	46	35	4	8	I	4	100
Planning and production engineers	I	35	38	4	19	2	I	100

NOTE: Figures are in percentages.   SOURCE: Questionnaire Survey

Without having lost their individual aspirations, many technicians have acquired a sense of collective dissatisfaction with their work and market position *as a group*. It was as a group member rather than as an individual that interviewees responded with comments such as 'generally the technical staff are in a poor position in the establishment, being considered equivalent to clerical staff, whose responsibilities and qualifications are in fact lower than those of any laboratory technician'. Or again: 'We are being drowned by highly paid and highly qualified graduate staff, and also by highly organized but poorly qualified manual workers who belong to powerful unions.' The ability of technicians to choose comparative reference groups such as clerks and manual workers had risen with the introduction of job evaluation and other formal grading and reward structures into the plants visited in the study. The value of this growing feeling of deprivation is well illustrated in Table 5.7 below.

In the process of judging their own needs it seemed difficult for the technicians to separate intrinsic factors from their need for a higher level of earnings or better working conditions. To some extent many status symbols such as not 'clocking-in' and

Table 5.7 Do you feel that the status of technicians is high enough in your company?

Occupational group	Definitely yes	Yes	Average	No	Definitely no	No reply	Total
Draughtsmen	1	12	33	41	10	3	100
Laboratory staff	3	13	26	38	18	3	100
Planners and production engineers	1	18	35	33	10	3	100
Quality-control technicians	2	11	29	46	10	3	100

NOTE:
Figures are in percentages.   SOURCE: Questionnaire Survey

the use of certain facilities had been culturally internalized. Advancement and increased responsibility in terms of the work role were seen as being a means to the attainment of these prized rewards. In general the role expectations of technicians were very closely structured by notions of what was an 'appropriate' compensatory level of extrinsic rewards. Upward mobility, whether through individual promotion or group progression was seen in relative terms and judged according to the rewards of others. It was also seen in terms of expected future rewards, principal among which was the prospect of promotion. It may well be that for many young laboratory assistants, as for the time-serving apprentice, present dissatisfaction with salary and other conditions was bearable so long as the cost could be discounted against future rewards. Many of the planning and production engineers might see themselves in a 'bridging role' between their initial post-training job and their anticipated managerial role.[1] They might also be prepared to discount present losses in salary against the expected intrinsic and extrinsic rewards that would go with a future position of higher authority.

It seems therefore that the possibility of judging a 'sufficiency' of 'extrinsic' factors in order to provide a floor for achievement-oriented needs may have to be related to a large number of dynamic factors. Since extrinsic rewards are attached to position

[1] For an elaboration of the use of this occupational typology see J. H. Smith, 'The Analysis of Labour Mobility' in B. C. Roberts and J. H. Smith (eds.), *Manpower Policy and Employment Trends* (L.S.E. – Bell, London 1966).

or status, and because advancement is demonstrated by changes in rewards, the 'hygiene' and 'motivating' factors cannot be treated as independent variables but as highly interdependent. Because the level of attainment to which the technician aspires at any given time is related to an ability to estimate future returns to his (universal) skills or to his (specific) company job, conflict will arise if it is not possible to achieve a dynamic career. There is a danger that employers mistake the temporary satisfaction of technicians with their terms of employment with a permanent situation, when in fact the low-paid young laboratory assistant is only discounting present costs against a 'bright future' and is not prepared to wait too long to see it attained[1] before seeking it elsewhere. Secondly, a satisfaction has to be seen as relative to the chosen reference groups of the technicians, and these may vary according to the orientation to work that is derived from home, education, training and other external factors; these may also vary according to place and circumstances, over time, and between individuals.

The orientation to work

In Chapter 2 it was suggested that the aspirations and expectations of technicians were highly structured by their external background. It seems likely that many of the frustrations of young laboratory assistants result from their entrance to a job which disappoints the hopes raised in grammar-school or technical-school education. They are frustrated in their attempts to gain higher technical qualifications through part-time study by the knowledge that yet another step has to be taken before they can surmount the graduate barrier in the promotion field. More immediately the tasks they are given are very often unsatisfying in the extreme, while the pressures placed upon them by early marriage give little incentive to regard this as a temporary phase to be overcome before higher earnings levels are attained. In this survey, a majority of laboratory assistants had been promoted from the shop floor, and it was among these people that the level of satisfaction with utilization of skills was highest.

Draughtsmen were fairly clearly divided into those who had

---

[1] For a more sophisticated explanation see R. A. Ulrich, 'A theoretical model of human behaviour in organizations: an eclectic approach', unpublished dissertation (London School of Economics 1969).

trained as apprentices, some of whom had the highest technical qualifications, and those who had been through a staff training scheme. The differences in attitudes expressed by these two groups were quite significant in a number of aspects. These are to be explored in Chapter 8. It is however possible to distinguish ex-apprentices in the drawing office as being among the most ambitious of the technicians: they were notable in that they retained their ambitions over a significantly longer period of their career, that is, after people of a similar age-group had become adjusted to their limited life chances. Ex-apprentices in all occupations set significantly greater value on the utilization of their skills and task-oriented needs. Quality controllers contained the highest number of older technicians and were significantly more likely to be staff trainees than other groups. They were also more likely to have moved from another firm, and in a number of cases it seemed clear that the move had also involved a promotion. The 'staff training' course was therefore some-times an induction to a new job and new status relatively late in their career. They were, however, the group that had adjusted best to their relatively limited opportunities for future promo-tion.

Production and planning engineers were a category contain-ing at least two dichotomous groups. The planning engineers tended to be ex-apprentices who were often formally well qualified, with previous experience as draughtsmen or junior technicians. The production engineers were less formally well qualified and had had a shorter period of training. This led to marked differences in their levels of aspirations.

Possibly because of their geographical separation from other groups within the place of work or possibly because of their purely 'staff' function, draughtsmen and laboratory assistants generally appeared to constitute two quite separate cultural entities within the work organization. Many differences were quite clearly attributable to differences in their orientation to work—their education, training and socialization determined to a large extent the groups to whom they referred themselves in terms of pay and promotion. Though the differences in age were not always great, the line technicians had adjusted their promo-tion aspirations and expectations to those the company could create in a more 'realistic' way than had 'staff' technicians. The laboratory staff and draughtsmen were less identified with the company, and the former group were not so keen to enter general management. Many identified strongly with their section head, who was a technologist or staff specialist. Similarly, draughtsmen regarded their chief draughtsman as someone having a technical competence and a more legitimate form of authority than more senior management. Even so, draughtsmen were more likely to feel 'treated as a grade' than as an individual. All three of the other major groups were very much more likely to feel as if they were treated as individuals. Again, this is difficult to understand without taking into account the high and persistent level of aspiration among draughtsmen. Their work, it is true, has been subjected to a great deal of rationalization, and for some the type of work they were doing required little of the knowledge and practice gained in a five-year apprenticeship. As a result the personal interviews with their section heads which most technicians had at regular intervals were extremely frustrating experiences. But this frustration was as much due to the fact that draughtsmen had hopes of obtaining promotion, as determined by the nature of their present work situation.

Communications

As one might expect from a stratum of employees which used management as a normative reference group, technicians placed an overwhelming importance on communications. The nature of their feelings with respect to the effectiveness of communications was influenced by the basic value-patterns perceived to be attached to 'management' in the abstract. There are difficulties about talking of management in this way even at this level, because 'management' for technicians in a small plant placed geographically hundreds of miles from administrative headquarters means something quite different from 'management' at another plant attached to headquarters. Technicians may see the technologists who run their plant as 'managers' in personal career and locally normative terms, yet in considering the power of those managers to take decisions which actually *affect* their current market position they may well be aware that real authority rests at a different level of management structure. The resulting ambivalence in attitudes does not of course require the magnitude of geographical separation between the two levels of management structure. The section head or departmental manager, who is the immediate personification of 'management' with whom the technicians so strongly identify and the administrative heads, who take crucial marketing and financial decisions, from which stem so much of the insecurity and frustration apparent in the technicians' response, may both be located on a single site, but the social distance between them may be very great. From the written-in comments and from interview material it was apparent that the use of personalized communications through section heads was a source of status for technicians but also gave rise to a completely cynical feeling that they were being manipulated by higher management. The statistical evidence supported this impression gained from interviews.

Management competence in technical matters was challenged by less than a third of our informants,[1] and communications upwards on technical suggestions appeared to be regarded as at least average by a majority of the sample; opinion of them was particularly high in the laboratories. Technicians' views on access to management and especially to the supervisor were also generally quite high, and laboratory assistants and planning and production engineers are again likely to be most satisfied in this respect. The level of consultation about changes in work practice undertaken by management appeared to be quite considerable: 60 per cent of informants in three occupational groups thought communications good or very good in this respect, while among the fourth, planning and production engineers, communications were rated highly by 65 per cent of the sample.

Only among quality controllers was a strong identification with a management outlook also accompanied by a poor opinion of management's human relations skills: 68 per cent of those quality controllers identifying with management thought them bad or very bad in their grasp of human relations. This was despite the fact that in common with other technicians a majority, 51 per cent, identifying with management, also felt they had a ready access to management.

Opinions only began to divide more evenly across all our

---

[1] See D. E. Weir and O. T. H. Mercer, *Report on S.S.R.C. Conference on Industrial Relations and Social Stratification,* mimeo (Social Science Research Council, Cambridge 1969), for an interpretation of the importance of the technical skills of managements as legitimating their basis of authority. See also J. A. Banks, *Industrial Participation, Theory and Practice, A Case Study* (Liverpool University Press, 1963).

informants when they were asked about communicating *griev-ances* upwards. More draughtsmen and quality controllers were, for example, likely to find them average or bad. Since this feeling was related to union membership it is explored more fully in Chapter 8. Opinions were universally average to very bad on the subject of their knowledge of promotion policies—this despite frequent meetings with section heads.

Communications on departmental prospects and upon changes in product were seen to be bad or very bad by the large majority of technicians—among draughtsmen only 14 per cent thought that they were at all good. Similarly, opinion was united on the poorness of communications on promotion prospects. It was clear that there was a division of feeling here which might be attributable to size and to the extended lines of communication in the large corporation. However, it was also the selectivity in the types of information released and held back that created a great deal of unease across all plants covered in the study. This unease often sprang from some historical experience such as the instant dismissal of seven well-qualified technologists by the management of one geographically remote research and development plant after instructions from central management to rationalize. The personally rewarding face-to-face contact with the supervisor and departmental or local manager was not productive of an exchange of information relating to anything above the immediate task environment. The future of the technician was discussed at such interviews in terms of his personal career development. His disquiet largely stemmed from market decisions taken out of context of the immediate work organization. In several plants visited during the course of this study the opinion was expressed that local management were unable to take any decisions 'without referring to Head Office'. This is an aspect of the way in which organizational 'size', as expressed in the geographically scattered multi-plant firm, which retains a highly centralized decision-taking procedure, affected the morale of employees.

The second important dimension of dissatisfaction with intra-organizational communications was that which existed among planning and production engineers, quality controllers and other line technicians. Among the latter in particular, opinion of management communications, whether on task-centred subjects or in respect of the market position, was bad and even very bad. Their access to management and supervisors was poor and their opinion of the human-relations skills of management low.

These groups are technicians who, over a whole range of technological environments, may be seen as being attached to line production and working to the directives of line management to a much greater extent than are laboratory assistants or draughtsmen. These latter groups are isolated from line management, generally working close to a departmental head who is a functional specialist. By contrast, the quality controller feels himself to be an adjunct to line management and one whose qualities are not truly appreciated. Instead, the production-oriented section heads give encouragement to line workers, who are the source of the production effort required to meet their goals, but regard quality controllers as a necessary evil. Quality controllers are in fact constraints upon the manager's ability to operate the system as he would like, and often have to be seen in this light. Hence communications on changes in product may be regarded as significant in power terms.[1]

The desire exhibited by line technicians for professional status (see Chapters 3 and 8) has, then, to be seen in terms of the added authority it gives in their internecine 'struggle' with line management. They are often more highly qualified in a formal sense than the line managers themselves, and in the more sophisticated industries, such as electronics, have built up a positive code of ethics relating to the need to improve the actual design of the product through improved testing.

The desire for status among these line technicians is also related to their work situation *vis-à-vis* shop-floor workers. Unlike their colleagues in the laboratories and drawing offices, they have few outward trappings to identify their status. In many plants they were treated like shop-floor workers in respect of canteen provisions and clocking in and out for work. Yet they were generally expected to exercise some degree of authority as management representatives in their day-to-day relationships with shop-floor workers. It is scarcely surprising, then, that they should attempt to put some measure of social distance between themselves and the latter in the form of the representative organization they choose.

[1] For a more sophisticated theoretical analysis of this work situation see D. Pugh, 'Role activation conflict: a study of Industrial Inspection', *American Sociological Review*, Vol. 31, No. 6 (December 1966).

F

Planning and production engineers do not appear to share the same sense of deprivation in all areas of their work and market experience, but they *are* engaged in a confrontation with line management, which is to be seen in terms of their relative status power. The role of planning engineer within the establishment usually consists of planning the progress of a product, from buying in the component parts to scheduling the date upon which it is despatched to the customer. Planning engineers are normally concerned with drawing up an ideal-type schedule for a particular line of production, but the implementation of such a schedule is under the control of the works or production manager to whom such engineers are normally responsible.

Work-study engineers (who have been subsumed within the former title in our statistical categorization) usually operate at the task level of this planning process. In this position they are exposed to organizational pressures at different levels and of different kinds. They have to meet not only the demands of the Works Manager and his subordinates in charge of workshops, but also the claims of shop-floor employees, for whom they represent a key to higher rewards or to smaller work-loads. This is of course particularly true in plants using payments-by-results schemes, of which there were a number in the survey.

The effect of organizational size

A number of studies have pointed to the effect of organizational size and bureaucratization as factors in creating an increased impersonality in management–labour relations, and have even suggested that increased size is directly associated with increased feelings of alienation from the organizations and from the task. This feeling has usually been measured in terms of a syndrome of 'powerlessness, meaningless of the task, isolation and self-estrangement'.[1] Although there has already been shown to be some evidence of what might be considered an alienative state among some groups of technicians, this was by no means similar to that observed among many manual workers, for example those in Professor Blauner's study.[2] Nor did these states of feeling appear to be associated in any direct way with variations

[1] R. Blauner, *Alienation and Freedom* (Chicago University Press, 1964)
[2] ibid.

in size[1] as such, although it is not suggested that size is not a significant factor.

The plants in which the study was conducted covered a range of sizes, from a small research and development establishment employing 800 people, to a large steel mill employing 16,000, with a majority in plants with between 1,500 and 2,500 employees.[2]

In order to gauge their orientation to the organizational features of their work we asked technicians what size of company and what size of department they preferred to work in. The highest proportion of respondents (40–42 per cent) preferred to work in a company of 1,000 employees, 20 per cent to 34 per cent preferred companies between 1,000–5,000 employees, and only 10 per cent to 15 per cent preferred companies of over 5,000 employees. In this distribution there was a tendency to select a company of the same size-order to the one in which the technician was already working. Even if one assumed that these findings represented an initial attraction to the company or plant on the grounds of size alone (a somewhat unlikely event) it is important to remember that much satisfaction was gained from interpersonal relationships in the work-group. In practice between 72 per cent (laboratory assistants) and 85 per cent (quality controllers) worked in groups of fewer than twenty members. It seemed therefore that the size of the department in which the technician worked might be more important than the size of the overall plant. The largest proportions of technicians, 64 per cent of quality controllers, 44 per cent of draughtsmen, and half of other groups, did in fact prefer to work in departments of a size similar to the one in which they were actually employed.

The effect of the work-group

The majority of all technicians worked in groups of ten or fewer (about 5 per cent working alone) and the level of work interaction appeared unaffected by the production technology of the firm. In respect of this dimension of 'bureaucratization' most tech-

[1] These findings are supported for draughtsmen by the more detailed work of G. K. Ingham, 'Organizational size, orientation to work and industrial behaviour' *Sociology*, 1 (1967), pp. 239–258.

[2] A comparison of these two work situations is to be found in J. V. Eason and R. Loveridge, 'Management Policy and Unionization among Technicians', in *Proceedings of an S.S.R.C. Conference on Social Stratification andIndustrial Relations* (Social Science Research Council, Cambridge 1969).

nicians can hardly be considered to have been reduced to the homogeneous mass. With the encouragement of top management, supervisors generally maintained positive relationships with each member of the work-groups under them. In at least one company comprising two plants in the study the development of group supervisory style was a central part of a 'neo-human relations' philosophy adopted by head office. The level of satisfaction with access to their supervisor was uniformly high even among trade-union members, with as many as 70 per cent of respondents (90 per cent of non-unionists) among laboratory assistants saying that it was good or very good; and nowhere did more than 5 per cent of the sample consider it to be bad. The amount of communications between management and labour appeared high even if the level of satisfaction derived appeared, to say the least, somewhat variable.

When tested against the desire of technicians to become part of management, or whether they saw themselves as already part of the management team, the size of the group and the access technicians had to their supervisor seemed to hold no consistent relationship. If anything technicians working alone, particularly planning and production engineers, seemed to have greatest aspirations to management positions. It has been demonstrated in the laboratory tests of industrial psychologists that the level of individual aspiration is highly influenced by the level of achievement appertaining within the membership group.[1] It is clear that while some technicians had made adjustments to their aspirational levels, to match their diminishing chances of promotion over the course of their career, most technicians were less influenced by their immediate work colleagues in the formulation of their ambitions than by the influence of their previous training and the presence of those who had 'made it' into managerial or supervisory posts.

The effect of plant technology

One of the most powerful influences on the behaviour of technicians in the work situation may be the technology itself. Two post-war schools in British academic thought have suggested that the technology of the work-place is a major structural impera-

[1] See for example Kurt Lewin, 'The Psychological Effect of Success and Failure', *Occupations*, XIV (1936), pp. 926–930.

tive in determining the nature of work-place organization and social relationships. Of these, the thought of the Liverpool School is perhaps the more deterministic and has been developed by the work of Professor Joan Woodward.[1] The second, that of the Tavistock Institute of Human Relations,[2] has been primarily concerned with possible modifications to technological and other features of the organization so as to permit increased economic effectiveness and improved work-satisfaction among employees.

Professor Woodward has suggested that particular forms of technology are related to certain product markets, and that taken together the market and technological requirements of a particular commodity or service determine what is the most appropriate, that is the most effective form of organization. That is to say that the classical axioms of 'scientific management' relating to increased role-specialization or formal 'job-descriptions' and rigorously defined 'spans of managerial control' need not and perhaps should not apply to all situations. Thus 'one-off' or 'small-batch' production implies a unique definition of each product and of the processes which go towards its production. The market is also a high-risk one and the customer's good will depends on every single order being of a high standard rather than only a proportion of the output meeting his requirements. Quality control is therefore of much greater importance than it is in simple line-inspection work, and indeed it is difficult to think in terms of the traditional line–staff dichotomy, or even of centralized decision-taking. The task of developing each new product is an all-involving one in which the whole organization is 'the achieving unit'.

As opposed to the 'flat' structure of authority found in small-batch production, the status hierarchies in Professor Woodward's other two forms of socio-technical systems[3] are high, and, at the

[1] J. Woodward, *Industrial Organization: Theory and Practice* (Oxford University Press, London 1965).

[2] For example, E. L. Trist *et al.*, *Organizational Choice* (Tavistock, London 1963), or of the same school, G. Friedmann, *The Anatomy of Work* (Heinemann, London 1961), Chapters III and IV.

[3] The term 'socio-technical system' stems from the work of the Tavistock Institute which pre-dates that of Professor Woodward. It is used here to emphasize the fact that, as mentioned in the last chapter, the systems being described are of a multi-variate kind in which the role structure of the organization and human interaction within it are not only *acted upon* but also help to determine the nature of the organization. The main emphasis in the Tavistock work has been upon the determinate nature of the human interaction, and is to that extent very different from that of Professor Woodward.

top decision-taking level, rather remote from the point of production. These she designates: (i) continuous-flow or process systems, and (ii) large-batch or mass-production systems. Of these the latter comes closest to providing an antithesis of the flexible and informal role structure and interpersonal sentiments produced in the small-batch system. Not only is the hierarchy of authority very high and communications extended, but inter-action is more rigidly constrained and more direct: it relates much more to the popular concept of a bureaucracy. The span of control is widest at the first-line level of supervision and the nature of the problems involved in 'feeding the line' is much more conducive to crisis and stress than in the other two forms of socio-technical systems. This provides at the supervisory level yet another contrast between mass-production and continuous-flow production. Although the form of organizational structure associated with the latter system provides the highest number of hierarchical levels of authority, the nature of both the production process and the market strategies in process industries makes for fewer short-term crises. By the same token the high degree of technological innovation incorporated within the capital equipment requires a high degree of understanding and ability to deal with what are essentially indirect and abstract problems as they arise. Communications tend therefore to be of a qualitatively different kind, and the point-of-production problems to require a more responsible approach to the work role and to develop a longer-term perspective of the work situation and the exigencies arising within it.

A number of studies carried out in various countries have contributed supportive evidence for Professor Woodward's findings. A study of the introduction of the research and development function into some Scottish electronics firms suggested that a form of 'democratic team-work' described by the researchers as 'organismic' was more appropriate for this form of production in which the market outcomes were extremely unpredictable than the bureaucratic or 'mechanistic' type which had evolved in traditionally stable markets.[1]

Evidence also extends to the form and nature of the attributes discovered within role-players in each of the systems—their

---

[1] T. Burns and G. M. Stalker, *The Management of Innovation* (Tavistock, London 1961).

qualifications and skills *vis-à-vis* the technology and their attitudes and styles of behaviour towards each other[1] and towards the organization as a whole. The drift of such findings suggests the existence of much higher levels of estrangement and isolation from the goals of the company among staff employed in mass-production technologies and a generally lower level of morale in these firms.

Most of the research into the effect of the product market and the technology used by an organization has been concentrated on the situation at the point of production. Only rarely is the analysis extended in any depth to the relationship of ancillary departments or staff functions to the interactive system in the production process. For this reason the typological frame suggested in these works was of somewhat limited usefulness in evaluating the results of this study. A detailed organizational analysis is not attempted in this report, but it is possible to put forward some observations which could prove of value in the modification and extension of research in this area.

It was apparent from the company records and from our interviews that the formal structures concurred very largely with Professor Woodward's findings. Within plants in which a continuous-flow technology constituted a major part of the production process (few had only one form of technology in use) authority tended to be centralized, and the social distance between senior and site management and the department in which technicians worked was considerably more than within plants using small-batch or unit technologies. This was not invariably so, however. In one large oil plant the drawing office was situated in the same building as senior management. Although there was little evidence to show that this visibly affected the attitudes of management (despite a company charter and education programme designed to stimulate a 'democratic' management style), there seemed little doubt that physical proximity to management was much appreciated by the young and ambitious draughtsmen. It does therefore appear possible that the degree of flexibility in locating staff departments—if only physically—may be greater than envisaged in

---

[1] See for example R. Blauner, op. cit.; A. Touraine *et al.*, *Workers' Attitudes to Technical Change* (O.E.C.D., Paris 1965), pp. 16–21, or for a more quantitative approach, J. R. Bright, 'Does Automation raise skill requirements?' *Harvard Business Review*, Vol. 36, No. 4 (1958), pp. 85–98.

most previous studies. It also suggests that sentiments might relate to something other than the work process. The draughts-men in Professor Woodward's study gained a shared status and derived some prestige as a group from the fact that their function was located at a main junction in a vector of work-processes. Although this might involve them in considerable stress, espec-ially in mass-production technologies, the degree of organiza-tional dependence upon their role in the production process which existed at all levels more than compensated them for the costs of periodic crises. Yet from an inter-plant comparison of findings, it appeared that in addition to the satisfaction derived from a sense of group identity and solidarity in the task environ-ment, the individual status-needs brought into the situation by draughtsmen may receive a good deal of satisfaction from their sheer physical proximity to senior management, that is, the locus of their individual ambition.

The variations in response to questions on job satisfaction, perceived needs, morale and status reproduced in Tables 5.1–4 did not always accord with what one might expect in the light of the above studies. In order to attempt a preliminary assess-ment of differences in attitudes created by a movement along Professor Woodward's continuum of market and technological variables, the scores for the above measurements of work senti-ments expressed by informants were compared company by company in the light of the product market in which they operated and the technology used by each. It was difficult to discover much systematic relationship between the socio-tech-nical systems (as expressed in these crude aggregate forms) and the attitudes of employees. Their occupational status was more generally considered to be very low indeed across all four plants using continuous-flow techniques: this might accord with largely routine jobs of many laboratory assistants and testers. Yet it was equally as low in the smallest of the research and develop-ment units included in the study, in which tasks were quite different. The general level of morale was abysmally low. The three plants recording the highest levels of morale were in three separate technology types, all situated in the north of England. Technicians in these plants displayed significantly higher levels of satisfaction with the people they worked with. Surprisingly, in view of the general lack of interest in fringe benefits, staff in two of these organizations also expressed a significantly higher level of satisfaction with such deferred rewards. Since these samples were all taken in areas in which opportunities for other employment of a similar status were relatively low, the evidence therefore suggests that the local labour-market situation may help to account for the feelings of these technicians.

Of equal interest were the significant differences in the level of morale and in the levels of satisfaction expressed within two plants belonging to a single company and responding to a similar expression of the formal goals of the company conveyed in their organizational charter and guides to behaviour. This was true of two pairs of organizations, one pair in a continuous-flow technology, the other in large-batch, both with one plant in the North-West and the other in the South-East. In both cases the opinions expressed on their status by Southern technicians were considerably lower than those of their Northern colleagues. Within these matched pairs of plants differences in the relative levels of satisfaction were greatest in organizations using continuous-flow technology, and were higher in the Northern plant than in the Southern plant along almost all dimensions. They were however equally low on the feeling of identity with the company expressed by technicians. In both these regions technicians were not very highly satisfied with the use to which their skills were being put, nor did they derive much satisfaction from their task-centred activities. The differences were largely to be found in their feelings towards their group status and individual career opportunities. It is difficult to find an immediate explanation for the differences between the two sets of plants within the systems themselves. Both worked under highly professional management, and in the continuous-flow plants were guided by a specially drafted charter of aims, in the drafting of which a human relations consultant had played a major part. The Northern plant employed over twice as many employees as the Southern one, 4,500 as against 1,936, and might therefore be expected to be a worse place in which to work in the technicians' own terms. The relatively low morale in both plants is almost certainly accounted for in terms of similar problems: problems which have an almost universal ring about them. Younger technicians were largely employed on routine testing jobs. Despite generous part-time facilities for study these concessions were cut off abruptly at the age of 22, and in fact few young married men had achieved more than O.N.C., though

a small proportion had H.N.C. by this stage. They were aware
that most higher-grade jobs were being filled by graduates. They
were also aware that after a recent productivity deal the consolidated guaranteed earnings of operatives were greater than their
own take-home pay.

One laboratory assistant in the Southern plant wrote in his
reply, 'Industry treats us like rubbish but shows greater respect
for the man who sweeps the laboratory floor because sweepers
are harder to recruit.' This was a perfectly realistic statement, as
the pay category into which sweepers fell was actually higher
than that of some laboratory technicians after a productivity
deal. It was clear that the respondent regarded his salary level
as a mark of status and that the proximity of the manual worker
within the socio-technical system stimulated this response.

Yet despite their relative deprivation in terms of earnings the
technicians in the Southern plant placed the increased use of
their skills as paramount in changes that they would most
like management to make. Management were aware of the problem of under-utilization and had in fact adopted a 'human
resource development' approach to future manpower planning.
However, the inter-relatedness between 'advancement' and
reward was underlined by the comment of another technician,
who alleged that management were trying to use them 'as cheap-
rate technologists' in a current 'job-enrichment' exercise that
was being carried on in the laboratory.

In attempting to explain the difference between the two plants
it seems possible that the general state of local labour-markets
created a difference in the degree to which meaningful external
comparisons could be made by technicians. The unsettling
influence of the earnings and changes in earnings paid to manual
labourers doing sub-contracting work on both sites provided
sufficient evidence of the importance of such external reference
groups to technicians in adjusting their own status and market
position. It seems possible to suggest that the comparative
reference groups in the North were more localized and were
regarded in a different manner to those perceived by the Home
Counties worker. The Northern technician was generally able to
find more favourable comparisons, particularly in terms of the
security of his job, than was his Southern counterpart.

One very significant difference between the two latter plants
was in the degree and strength of identification with shop-floor

workers to be found among the technicians in the Southern plant. No fewer than 15 per cent of this latter group felt that their outlook was definitely in common with shop-floor workers as against only 2 per cent in the Northern plant. The next highest level of support in any plant was only 4 per cent in two Southern small-batch production plants, and the general level of identification with management did not greatly depart from the magnitudes shown in Table 5.5. The deviancy in outlook from the management norm which the technicians in the Southern process firm presented did not mean that they were not ambitious: only two other plants were more keen for promotion. *No* other group felt so strongly as this one that they had *not* been given sufficient opportunity for individual advancement: 75 per cent felt that there was insufficient opportunity for promotion, as against 64 per cent in the Northern plant. It seems likely that in the embryonic state of 'collective' consciousness displayed by these frustrated technicians the militant tactics of the unions representing manual workers, as against their own apparently ineffective staff association, provided them with an alternative strategy and reference group for action. At the other end of the spectrum those plants in which technicians displayed the highest amount of satisfaction in their work interests and in their status also displayed the highest sense of identity with management, despite their dissatisfaction with their salary and irrespective of the form of technology in use and the market situation.

It may not be without significance that in Table 5.8 the forms of communication that are found to be most unsatisfactory, after that concerning promotions, are those relating to changes in the product market and to departmental policy. The fact that behavioural analysts, whose work has provided the theoretical frame for this chapter, have concentrated so fixedly on the task environment, even while perceiving the market to be an important structuring feature of that environment, is to be regarded as a major weakness in their work. This is particularly so where, as in most of the plants visited in this study, the local management are no more than a link in the chain of market decision-taking which begins at the head of a multi-plant firm.

In discussing individual needs for 'security', for instance, many behaviourists have confined their suggestions to task-design and interaction within the place of work, yet it was clear, both from the disquiet expressed in interviews and from written

Table 5.8   Views of technicians on communications with management categorized by type of production technology

Subject matter of communications	Production technology		
	Unit/small-batch	Large-batch	Continuous-flow
On technical instructions about their work	56	66	61
On promotion	43	47	47
On changes in product which might affect their job	45	51	38
On their department's prospects	47	54	50
On communicating their suggestions upwards	60	72	69
On getting their grievances dealt with	53	61	61

NOTE:
The scores shown in this table are calculated   SOURCE: Questionnaire Survey.
from a five-point scale ranging from 'very
good' to 'very bad'.

comments on questionnaires, that informants were aware of the implications of the cancellation of a contract for the firm and of the consequent danger of a drop in earnings due to rationalization and redundancy. In their assessment of management communications it is therefore not surprising that a rather higher opinion is held of management's ability to convey and to receive information about subjects over which technicians have some direct control. A rather less favourable opinion is expressed on communications dealing with promotion, product changes and their department's prospects. This was particularly found in plants with a 'continuous-flow' technology, where the lines of communication were inevitably extended, but even within the more closely integrated structure of the 'small-batch' plant, communications on these subjects were regarded as very poor.

Most of the small-batch firms in this study relied heavily on Government contracts, and management interviewed in the study often expressed the view that their research and development function would be the first to be 'axed' in any major company retractions. It is therefore hardly surprising that technicians' were also apprehensive, and that their anxiety should be displayed in attitudes to other facets of their work and working conditions.

Professor Woodward's thesis suggests that the uncertainty of product markets open to small-batch unit production techniques is one of the causative factors in the formation of organismic groups. However, the existence of shared decision-taking at plant level does not necessarily offer any degree of security to employees—or managers—working in multi-plant firms. Since most of the plants included in our study were of this kind, relationships within the work situation appeared to have little direct effect on the way technicians viewed their market situation, whatever the technology employed.

The work environment and technological change

There appear to be two important facets to the influence of the technological environment on technicians' attitudes. One relates to the satisfactions they derive from the utilization of their skills and their previous fund of knowledge, the other to the manner in which they refer to other groups in determining what are their 'just' rewards. This latter may, as we have suggested earlier, be related to some previously obtained qualification or 'objective' level of achievement, but since aspirations were high even among the most lowly qualified it was clear that their career goals were derived by reference to the standards set by others, in particular by the technologists or middle management with whom they worked. To the extent that they saw their hopes of promotion as being justified they retained them even in the face of constantly reducing promotion opportunities. Yet to the extent that these highly ambitious white-collar employees were brought into contact with manual workers within the place of work their desire to differentiate their status as a group was considerably heightened.[1] The transition from individual dissatisfaction with some aspect of their work which was regarded as intrinsically important, e.g. utilization of skills, was also translated into group terms through comparison with the work of others, for instance floor sweepers.

The comparisons became particularly crucial in periods of technological or social change within the work situation which led to unfavourable comparisons between the status of technician groups and manual workers. This was particularly likely

[1] The model being developed here was first set out in an article by David Silverman, 'Clerical ideologies – A research note', *Sociology*, June 1968, pp. 327–333.

to occur in situations in which manual groups acquired new responsibilities, or more often simply when their earnings were consolidated in a guaranteed salary structure. On such occasions technicians were most likely to express their complaints, not simply in terms of their aggrievement at their new market position, but also at the nature of their tasks. This happened in one steel strip-mill. The change in relative earnings of laboratory assistants, who had been exercising some supervisory responsibility in taking samples in extremely difficult and onerous circumstances, became the catalyst for long-held feelings of resentment against management for treating them (in their view) as if they were manual workers. The bitterness engendered resulted in a major breakaway of these technicians from the industrial (largely manual) union of which they were then members.

But within the same plant the trend in the task content of the technicians' work was by no means only in one direction. For example, electricians had acquired totally new skills and responsibilities within the maintenance and adjustment of the computerized control systems operating the flow of production. In order to do so, however, they had had to demonstrate their ability to master new skills and to cope with new levels of abstraction demanded in their changed tasks. Older technicians found some difficulty in adjustment; such personal adjustment problems were not wholly confined to older people. To the aspiring but less able younger man it was clear that these personal problems were being expressed in aggression over a wide range of issues.

Whatever the direction of change it was clear that technicians had a notion of what was the appropriate base point from which to start any assessment of earnings. Added responsibility and new skills were seen as deserving higher financial rewards. Diminished responsibility or dilution of tasks were resisted *per se,* but were evidently linked to a desire to maintain their earnings differential. But underlying their in-work comparisons and their view of their position in the local status hierarchy was an awareness of changes in their wider market situation, of which the increasing standard of education required of a potential technologist or manager was most important.

In order to explain differences in attitudes between plants it has been necessary to take into account the needs brought into

ORIENTATION TO WORK AND THE WORK SITUATION 167

the organization by technicians having regard for their previous experiences. In so far as these represented a continuity of similar work roles or a similarity in terms of career progression, there appeared to be a generality of aspirations and expectations which allowed for a group identity of feeling, not only in terms of any existing and individualized work situation but also in a collective sense. In the development of such feelings of 'occupational consciousness' the major reference points were to be found in and around the work situation, while underlying all such comparisons was the market situation of the group and its position in relation to others whom it regarded as significant. In so far as the socio-technical system of the plant allowed such comparisons to be made and gave rise to comparisons that were seen to be 'inequitable' by the technician group, it was extremely significant. This did not, however, prevent some managements from being relatively successful in meeting the needs of technicians, whatever the technological context or the local labour-market situation. This is a theme that will be taken up and developed in Chapter 9.

6 Salaries and Conditions of Employment

The Department of Employment collects and publishes figures of administrative, technical and clerical employees' earnings, but only for certain industries and services; these data are not broken down into individual occupations. Further data are available on technical staffs' salaries in three reports of the National Board for Prices and Incomes.[1] The Prices and Incomes Board data, based on a survey undertaken in June 1967, are confined to the engineering collective bargaining unit. To secure a more relevant picture of the earnings of the technicians covered in this study, the firms participating were asked to provide figures for the earnings trends of their technical staff over the previous five years. Only three firms found it possible to provide comprehensive figures: as a result of these statistical limitations this study of the structure of the labour market for technical staffs is inevitably restricted.

The Survey results

In nine of the firms surveyed the graded salary-structure had been established unilaterally by the company; in some cases

[1] National Board for Prices and Incomes, Report No. 49, *Pay and Conditions of Service of Engineering Workers* (1st Report), Cmnd 3495 (H.M.S.O., London, December 1967).

National Board for Prices and Incomes, Report No. 49, *Pay and Conditions of Service of Engineering Workers* (Statistical Supplement), Cmnd. 3495-1 (H.M.S.O., London, 1967).

National Board for Prices and Incomes, Report No. 68, *Agreements made between certain Engineering Firms and DATA*, Cmnd. 3632 (H.M.S.O., London, May 1968).

it was based on job evaluation. Within each grade there was a series of bands, and depending upon his performance, which was reviewed by the firm annually, the technician moved up each band and grade. Increments were generally not automatic, and at one firm complaints were heard of the incomes policy being used as an excuse not to pay incremental increases.

In other firms there was regular collective bargaining, the extent of which varied. In oil refining the companies did not belong to an Employers' Association, but were involved in plant bargaining with some of their technicians. In this case laboratory assistants were members of a union which had been accorded plant negotiating rights. Some other technicians in the plants were members of unions, but the company did not consider the degree of unionization sufficient to concede negotiating rights. In iron and steel, the establishments surveyed had conceded negotiating rights to DATA and to technicians in the chemical branch of the British Iron and Steel Kindred Trades Association. However, the companies had recently been nationalized, and although the British Steel Corporation had requested that all the white-collar staff should join a trade union, there was a dispute as to which union should be recognized for this purpose. Some of the unorganized technicians wanted to join the A.S.T.M.S., and a jurisdictional dispute between this union and some of the manual-worker unions with a white-collar section developed.[1]

General level of salaries

Empirical evidence shows that salaries have been declining relative to wages.[2] D. Seers has provided estimates of changes in the average level of salaries and wages between 1938 and 1949. His calculations suggest that wages rose by 136 per cent in the period but salaries by only 72 per cent. In the post-1948 period the number of salary-earners in manufacturing industry has continued to increase at a faster rate than that of wage-earners, so that the total salary bill has grown more than the total wage bill. However, the average salary has not grown as much as the average wage, reflecting in part the increase in the number

[1] See the *Report of a Court Inquiry under Lord Pearson into the dispute between the B.S.C. and certain of their employees*, Cmnd. 3754 (H.M.S.O., London, August 1968).

[2] D. Seers, *Levelling of Incomes since 1938* (Blackwell, Oxford 1951).

of salaried workers at the lower end of the salary range. It is
possible that the average salary of technicians has followed a
different course from the average of all salaries in manufacturing
industry.

Table 6.1   Wages and salaries in manufacturing industry since 1948

Year	Total wage bill	Total salary bill	Estimated average annual wage	Estimated average annual salary
1948	100	100	100	100
1952	135	147	130	123
1956	187	209	172	149
1960	225	284	212	184
1964	269	372	257	223
1965	306	448	293	249

SOURCE: National Income and Expenditure.

Table 6.2   Distribution of technical staff salaries

Salary	Number	Percentage	Cumulative percentage
Less than £900	86	9.4	9.4
£900–£999	68	7.2	16.6
£1,000–£1,099	107	11.3	27.9
£1,100–£1,199	98	9.6	37.5
£1,200–£1,299	138	14.7	52.2
£1,300–£1,399	117	13.0	65.2
£1,400–£1,499	78	8.3	73.5
£1,500–£1,599	84	9.2	82.7
£1,600 and more	166	18.3	100.0
Total	942	100.0	

SOURCE: Questionnaire Survey.

The extent to which the general compression of salary earnings
relative to wages applies to technicians is a matter of some
significance to their behaviour. Technicians' perceptions about
the relative movements of their earnings suggested the hypothesis that there had been a narrowing of differentials with
wages, and this was a factor explaining the behaviour patterns

of this occupational group, for example, their attempts to raise their earnings by moving to other firms and occupations, or joining a trade union. To throw light on the situation respondents were asked to state their salary grade. The results are shown in the table above.

The general level of technicians' salaries

When salary levels were examined by region it was found that they were higher in the South-East and North-West than in the other two regions. The labour markets for technicians' occupations were tighter in the South-East and North-West regions than in the South-West and Northern regions, and this was undoubtedly the factor mainly responsible. However, in the North-West a large proportion of the sample was accounted for by a high-paying, highly productive chemical firm, which may have overstated the true position in that region. The South-East results may have been under-estimated owing to the greater amounts of contract draughting in this area compared with the North-West. Contract draughtsmen were excluded from the study because of their non-appearance on firms' Industrial Training Boards' returns.

Table 6.3   Distribution of technical staff salaries by region[1]

Region	Less than £900	£900 –999	£1,000 –1,099	£1,100 –1,199	£1,200 –1,299	£1,300 –1,399	£1,400 –1,499	£1,500 –1,599	Over £1,600	Total
South-east	4	6	10	12	16	16	9	11	18	100
North-west	15	6	11	8	11	8	9	9	24	100
Northern	13	13	15	15	30	11	2	0	0	100
South-West	16	15	21	10	13	10	8	4	5	100
Overall Total	9	7	11	10	15	12	8	9	18	100

(1) Standard regions of the Department of Employment and Productivity.   SOURCE: Questionnaire Survey

The spread of earnings among technicians was greatest in the North-West and lowest in the Northern region. The dispersion of earnings in the South-East and South-West regions were similar, though levels were higher in London and the South-East. The age distribution of the technicians of the sample in the respective regions might have influenced the results, but the general position probably remained little affected.

Occupational differentials

There is an abundance of literature on wage structures concerning manual workers, but little on white-collar workers.[1] Questionnaire survey returns enabled a technician's wage-structure to be constructed at a point in time. Unfortunately information was not available to permit an analysis of past movements of pay levels for different groups. However, even in the absence of any past similar study, the survey may provide a bench-mark for future studies of technician wage-differentials.

There were significant differences in the median level of salaries of the broad occupational groups of technicians. Table 6.3 shows that 70 per cent of laboratory staffs, 92 per cent of the draughtsmen, 95 per cent of planning and production engineers and 73 per cent of quality-control technicians were earning more than £1,000 at the time of the study. The corresponding figures for those earning over £1,600 were: laboratory staffs 14 per cent, draughtsmen 24 per cent, planning and production engineers 13 per cent, and quality-control technicians 9 per cent. Of the four occupational categories the highest salaries were received by draughtsmen, followed by planning and production engineers, laboratory staffs, and quality-control technicians.

The spread of earnings within the occupational groups was greatest among laboratory staff, with a mean quartile deviation[2] of £249 and a coefficient of quartile deviation of 0.21. The dispersion among quality-control technicians was less than that among the laboratory assistants, but greater than among draughtsmen and planning, and production engineers.

(1) *Industry and regional differentials*

The differences between the occupational groups are analysed at the industry and regional level in Tables 6.5 and 6.6. Draughtsmen received the highest salaries in chemicals, where

[1] One exception is M. Fogarty, 'The White-Collar Wage Structure', *Economic Journal*, 1957.

[2] The best measure of the spread of earnings is the standard deviation, but this measure could not be used because there was an open class interval of £1,600 and over. It was not possible to obtain any guidance as to the range of salary in this class as the variation within it was very large. Thus the class mark (i.e. the class mid-point) could not be reliably estimated. Therefore the less useful measure of the mean quartile deviation and the co-efficient of quartile deviation, i.e. $\dfrac{Q_1 - Q_1}{Q_1}$ were used.

75 per cent earned over £1,570 and 10 per cent £1,690 or more. In metal manufacture the lower quartile was only £975 and the highest decile £1,300. At the median and upper quartile levels the salaries of draughtsmen employed at vehicles firms appeared to be above those in engineering. The highest salaries occurred in chemicals, which is a new and expanding industry and demands draughtsmen who are able to help with the planning and design of new plant. One of the companies in the sample was about to undertake an expansion programme that would require a large number of draughtsmen and designers. In order to obtain sufficient draughtsmen, chemicals firms were having to pay high salaries to bid draughtsmen from neighbouring firms.

Table 6.4 Spread of technician salaries – four occupational groups

Occupational Group	Lowest quartile	Median	Upper quartile	Highest decile	Mean quartile deviation	Coefficient of quartile deviation
	£	£	£	£	£	
Laboratory staff	914	1.132	1.141	1.167	249	.21
Draughtsmen	1,203	1,352	1,590	1,660	149	.14
Planning and production engineers	1,100	1,200	1,470	1.620	185	.14
Quality controllers	983	1,111	1,386	1.512	202	.17

SOURCE: Questionnaire Survey.

The state of the product market explains the low salaries in metal manufacture. At the time of the survey a world surplus of steel meant that the demand for draughtsmen for the planning and designing of new developments was low, since expansion was not envisaged. Moreover, one of the firms in the industry was in a somewhat monopsonistic position and was able to satisfy its demands for draughtsmen without having to compete strongly with other employers.

The salaries in vehicles and engineering were similar at the upper end, but at the lower end engineering draughtsmen's salaries appeared higher than those of vehicle draughtsmen. These two groups are covered by the same collective bargaining unit, namely the Engineering Employers' Federation and the

Confederation of Shipbuilding and Engineering Unions, which establishes minimum scales for DATA members employed in federated firms. However, there was some additional market effect on the level of salaries in that vehicles firms in the survey were expanding and experiencing recruitment difficulties. The three firms engaged in 'bidding' draughtsmen away from neighbouring firms, were aware they were giving an upward twist to salary levels. The engineering firms did not appear to have the same supply-deficiency problem, since they filled draughting vacancies by recruiting from the shop floor. The spread of earnings of draughtsmen within industries was greatest in vehicles, where there was a mean quartile deviation of £245, and least in chemicals, where it was £50.

The range of salaries for 75 per cent of the laboratory staff in chemicals was between £900 and £1,560, in metal manufacture £1,080–£1,260, in vehicles £950–£1,100 and in engineering £970 –£1,370. In both chemicals and engineering 10 per cent were earning over £1,599. These differences in the structure of salaries reflected a number of factors, for example, age distribution within the industrial order, differential rates of productivity for the industry as a whole, and the supply and demand for the occupation in the industry. The dispersion of salaries for laboratory technicians in vehicles was low in relation to the other industrial orders because the demand for laboratory staff in this industry was relatively low. This group of technicians was earning slightly less in engineering, where the dispersion of earnings was greater than in vehicles. The salaries of planning and production engineers employed in vehicles appeared to be the highest, with 50 per cent earning over £1,370 and 10 per cent over £1,670.

Quality-control technicians appeared to receive the highest salaries in the chemical industry, a situation due to the fact that the type of testing and inspecting done by quality control in chemicals required higher skills (where skills were defined on the basis of education) and was of a less routine nature than that done in vehicle assembly and engineering. In chemicals the inspectors had the responsibility of sanctioning the re-opening of plant worth millions of pounds after a shut-down. The scarcity of higher skills gave inspectors in chemicals a higher market rate than quality controllers in other industries.

Table 6.5 Dispersion of technician salaries for occupational group by industrial order[1]

	Lower quartile	Median	Upper quartile	Highest decile	Mean quartile deviation
Draughtsmen	£	£	£	£	£
Chemicals	1,570	1,630	1,660	1,690	50
Metal Manufacturing	975	1,170	1,260	1,300	140
Vehicles	1,080	1,400	1,570	1,640	245
Engineering	1,170	1,290	1,510	1,640	170
Laboratory staff					
Chemicals	900	1,200	1,560	1,650	330
Metal manufacturing	1,080	1,140	1,260	1,325	90
Vehicles	950	1,000	1,100	1,150	75
Engineering	970	1,140	1,370	1,599	200
Planning and production					
Chemicals	—	—	—	—	—
Metal manufacturing	—	—	—	—	—
Vehicles	1,240	1,370	1,520	1,670	140
Engineering	1,199	1,275	1,599	1,640	200
Quality-control technicians					
Chemicals	1,399	1,599	1,650	1,690	125
Metal manufacturing	—	—	—	—	—
Vehicles	1,125	1,199	1,299	1,350	87
Engineering	957	1,060	1,220	1,390	132

NOTE:
(1) Industry: based on the Standard Indus-   SOURCE: Questionnaire Survey.
trial Classification (1958).

The North-West appeared from the survey as the area with the highest salaries for draughtsmen, with 75 per cent earning above £1,260, 50 per cent above £1,460 and 25 per cent above £1,632. In the South-East region draughtsmen's salaries were slightly lower. The sample of firms in both regions included a high-paying, highly productive oil refinery, and two medium-sized electronics firms, as well as a small firm engaged in specialized tasks. However, the generally lower level of the salaries of draughtsmen in the Northern and South-West regions was probably an accurate reflection of the state of the technician labour-market in these areas.

Amongst laboratory staffs the highest salaries were found in the South-East. An explanation of the fact that at the lower end of the scale salaries of laboratory staff in the North-West were lower than those of the North may be found in the large numbers of young laboratory assistants in the returns for the North-West. This was also reflected in the wide dispersion of laboratory staff earnings in the North-West. The higher upper salaries of the North-West in relation to both the North and the South-West reflect the better market position of its laboratory staffs.

Table 6.6   Dispersion of technician salaries for occupational group, by standard region

Region[1]	Lower quartile	Median	Upper quartile	Highest decile	Mean quartile deviation
	£	£	£	£	
Draughtsmen					
South-East	1,233	1,410	1,605	1,660	186
North-West	1,260	1,460	1,632	1,680	186
Northern	975	1,133	1,210	1,299	118
South-West	1,099	1,240	1,290	1,599	95
Laboratory staff					
South-East	1,040	1,220	1,580	1,640	270
North-West	900	1,125	1,510	1,648	305
Northern	1,020	1,180	1,270	1,350	125
South-West	900	999	1,099	1,199	100
Planning and production					
South-East	1,205	1,280	1,440	1,620	118
North-West	1,099	1,270	1,399	1,640	150
Northern	—	—	—	—	—
South-West	1,150	1,330	1,425	1,625	138
Quality-control technicians					
South-East	980	1,113	1,299	1,499	160
North-West	1,050	1,360	1,610	1,670	280
Northern	—	—	—	—	—
South-West	930	1,005	1,060	1,125	65

NOTE:
(1) Standard Regions as revised by the   SOURCE: Questionnaire Survey.
    Ministry of Labour, 1966.

Amongst planning and production engineers levels of salaries were higher in the South-East region. Some 75 per cent were earning more than £1,205, 50 per cent £1,280 and 75 per cent £1,440. South-West salaries appeared to compare favourably with the North-West and South-East. This situation was probably due to the high compensation that a company in this region had to pay to overcome the disadvantage of working in an isolated area. It was difficult to attract planning and production engineers into this region, and a high wage had to be paid to induce them to come.

Quality-control technicians' salaries were highest and spread greatest in the North-West. In the South-East salaries were slightly lower than in the North-West, but the spread was smaller.

(2) The technician—manual worker differential

As data on manual workers' earnings were not collected from the firms surveyed a quantitative approach to the technician–manual worker differential was not possible. To obtain some statistical comparisons, an approach was made to personnel departments of the companies participating in the study. Three firms returned usable replies concerning the movement of technicians' salaries in relation to those of skilled manual workers. Two of the returns showed that over the previous five years the technician–manual worker differential had widened, but in the third it had narrowed.

An engineering firm in London made a comparison between the movement of the basic salaries of thirty craftsmen on weekly staff and those of thirty technicians, all of whom had been employed in the company over the previous five years. The result of this comparison was that the movement for technicians was of the order of £380–£400 in five years, and for skilled men £200–£250. In both absolute and relative terms, in this firm, the salary differential had increased in favour of the technician. Another engineering firm in Lancashire provided data showing a similar trend. In the period January 1962 to September 1968 the salaries of their technicians increased by $45\frac{1}{2}$ per cent, those of skilled maintenance fitters 17 per cent, weekly staff male employees $43\frac{1}{2}$ per cent, and hourly paid male employees on piece-rates $27\frac{1}{2}$ per cent. An aircraft firm in the South-East, however, showed a narrowing of this technician–manual worker differential. During the period December 1963 to November 1967 the wages of skilled fitters increased by 19 per cent but the salaries of draughtsmen and technicians by only 15 per cent.

The way in which technicians perceive movements in relativities are important in their behaviour. Table 6.7 shows the technical staff's replies to the question: 'Whether or not you feel you are being paid enough often depends on what others are paid. With which of these following groups would you compare yourself, when deciding whether your pay is fair?' Fifty-five per cent answered that other technical staff were in their 'orbit of coercive comparison', 19 per cent answered 'supervisors', and 22 per cent 'manual workers'. This suggests that although the technician–manual worker differential was of concern to technical staffs their concern with pay comparisons was greater in relation to workers doing jobs similar to their own. However, the fieldwork aspect of this research revealed some harsh comments about technicians' pay in relation to that of the shop floor.

Table 6.7 Pay reference groups of technical staff

Occupational group	Other technical staffs	Super-visors	Skilled manual workers	Unskilled manual workers	Semi-skilled manual workers	The pro-fessions	Others	Total
Laboratory staff	39	20	29	3	1	1	3	100
Draughtsmen	60	16	19	3	1	0	3	100
Planning and production engineers	67	18	10	0	1	0	3	100
Quality controllers	54	18	17	3	7	0	2	100

NOTE: Figures are in percentages rounded to the nearest whole number.   SOURCE: Questionnaire Survey

At an engineering firm in London a draughtsman remarked that most of the people in his office felt that the technicians' wage structure in the company was wrong.

> The differential between the toolroom and the drawing office is negligible and the position worsening. An apprenticeship was desirable ten or fifteen years ago but today it is not. There is no future in Mechanical Engineering today when one goes to night school, studies hard, becomes qualified and is then not financially rewarded. The effort does not seem worth while.

In an aircraft firm in the South-East draughtsmen were angry about the shop-floor differential, and complained that the traditional differentials had been reversed and should be restored. Although an engineering firm in Lancashire provided data to show that the differential with the shop-floor worker had widened in favour of technicians, in interviews at the plant draughtsmen complained that the pay differentials between draughtsmen and shop-floor workers appeared to be narrowing. Their concern centred around differential rates of overtime between the shop-floor and technical staffs. At a company involved in metal manufacture, interviews revealed a great deal of ignorance on the part of draughtsmen about the amount of earnings on the shop floor. Nevertheless the questionnaire survey returns show that 91 per cent of draughtsmen considered the pay of technical staff insufficient compared with that of the shop floor.

Testers and inspectors in electronics firms in London complained that they were were underpaid compared with shop-floor workers in earnings rather than basic rates. Shop-floor workers were seen as having powerful unions pressing their case. It was felt that testers and inspectors should be paid more and that interest in the job was being bought at the cost of lower earnings. The group considered management did not feel obliged to pay high rates to non-production workers; for example, weekly staff were paid a flat rate for overtime. The responsibility of the work, they argued, was not reflected in the pay differentials. One tester contended that in his job if a baby received a minute overdose of X-rays it could be killed, and it was his responsibility to see that this did not occur. Shop-floor workers had no such responsibility.

Laboratory staff in chemical firms strongly criticized the differential between their own salaries and the wages of workers in the plant. In an oil refinery in the North-West, a laboratory technician complained: 'It is a source of discontent that craftsmen who were considered unsuitable for laboratory work are able to earn more money than laboratory technicians. It is a waste of time studying to gain a qualification if at the end of the course the pay differential over non-qualified is insufficient.' Planning and production engineers at various companies complained bitterly of narrowing differentials in relation to the pay of floor workers. At a vehicles firm one stated that he and his colleagues considered salaries too low, particularly in relation to the earnings of the piece-rate workers. In their opinion slack piece-work rates were largely responsible for the inconsistencies of salary structure in the company.

This narrowing of differentials was seen not so much in terms of basic rates but of earnings. Frequent references were made to the amount of overtime payment enjoyed by manual workers. Technicians were expected to take time off in lieu of overtime worked, and where it was paid it was at a flat rate and not at a premium rate, as for manual workers. The complaint concerning differential overtime payments has to be judged against the background of information collected in the survey on overtime working. In the capital-intensive chemical firms, the growth of productivity bargaining was leading to the elimination of the necessity for overtime working by technical staff. The creation of greater job and time flexibility meant that staff could be switched between sections within departments or possibly between departments, and the normal numbers of working hours could be worked at varying times of the day or week. In return for flexible working, a disturbance payment was made.

Table 6.8 shows the frequency of overtime worked by technicians. There were considerable variations between the broad occupational groups. More than half of the draughtsmen worked little overtime and nearly 67 per cent of planning and production engineers rarely or never worked overtime, but quality-control technicians, service engineers and supervisory grades appeared to work a good deal of overtime compared with the other technician groups.

Table 6.8 Do you often work overtime?

Occupational group	Very often	Often	Not often	Rarely	Never	No answer	Total
Laboratory staff	11	12	25	32	19	1	100
Draughtsmen	3	10	29	37	20	1	100
Planning and production engineers	4	10	20	48	19	0	100
Quality-control technicians	22	20	28	21	6	3	100

NOTE:

Figures are in percentages.   SOURCE: Questionnaire Survey.

When overtime was analysed by industry, draughtsmen, except for those in vehicle firms, worked negligible overtime. Laboratory staff in chemicals and metal manufacture worked little overtime, but in vehicles 66⅔ per cent worked overtime very often or often, as opposed to 33⅓ per cent who rarely or never worked overtime. Planning engineers rarely or never worked overtime in the engineering and electrical goods firms, but in vehicles 37 per cent answered they often or very often worked overtime. In all the industrial groups the percentage of quality-control technicians who answered that overtime was often or very often worked was higher than that of those who replied that they rarely or never worked overtime.

Although some overtime was worked by technicians, there was little evidence of the regular institutional overtime worked by manual workers, except in the case of three groups mentioned.

Factors influencing attitudes to manual-worker differentials

The attitude of technicians to the shop-floor differential may have been influenced by many factors, for example trade-union membership, the level of qualifications and level of salary. However, the survey revealed high dissatisfaction with salary differentials among all technical staffs, whether members of a union or not (see Table 6.9). Trade-union membership, then, was not a

Table 6.9 Attitude of technician staff by trade-union membership to the sufficiency of pay compared with shop-floor workers

Are you a paid-up member of a trade union for technical staff?	Do you think that the pay of technical staff is insufficient, compared with that of shop-floor workers?					
	Definitely yes	Yes	Not sure	No	Definitely No	Total
(a) Draughtsmen						
Yes	60	27	5	5	3	100
No	52	38	5	5	1	100
Total	55	33	5	5	2	100
(b) Laboratory staff						
Yes	70	22	2	5	1	100
No	57	31	2	7	1	100
Total	65	26	3	6	1	100
(c) Planning and production engineers						
Yes	40	20	0	20	0	100
No	37	43	2	18	0	100
Total	38	39	1	21	0	100
(d) Quality-control technicians						
Yes	46	48	0	45	2	100
No	52	25	5	12	7	100
Total	50	33	3	9	5	100

NOTE:
Figures are in percentages.   SOURCE: Questionnaire Survey.

significant variable in determining the attitude of technicians to the shop-floor differential, except among quality controllers, where it was only marginal.

Draughtsmen were strongly influenced in their attitudes to differentials by the feeling that they were inadequate as a reward for their level of qualifications. Questionnaire returns from draughtsmen suggested that their dissatisfaction with the differential was strongest at the top and bottom of the salary levels, and for one group in the middle (see Table 6.10). However, tests of significance showed that for draughtsmen the level of salary was not significant in influencing attitudes to shop-floor differentials.

When the qualifications of draughtsmen were listed in descending order of educational attainment, dissatisfaction was seen to be greatest among the most highly qualified (that is, degree, H.N.D. or H.N.C. standard). This was probably because these groups considered that insufficient payment was given to compensate for studying instead of enjoying leisure time. On the other hand, dissatisfaction was also very high among the least qualified, probably because these considered white-collar occupations to be higher than manual occupations in the companies' hierarchy of skills and therefore thought that the rewards there should be greater in order to encourage individuals to better themselves.

For laboratory staffs the pattern was not as clear as for draughtsmen. Dissatisfaction with the differential was apparent at all salary levels and was not concentrated in any extreme ranges. There appeared to be little direct relationship between attitudes to the differential and level of salary. When attitudes were studied in relation to qualifications, although the greatest dissatisfaction was to be found among the lowest qualified (i.e. those with no formal qualification and low grades of City and Guilds), unlike the draughtsmen, dissatisfaction among the most highly qualified was not as strong. Only 50 per cent of those with H.N.D. considered the shop-floor differential insufficient, while 20 per cent thought it to be sufficient.

The returns from planning and production engineers also showed a feeling of dissatisfaction with shop-floor differentials at all levels of salary. However, feeling about the insufficiency of the differential did not appear to be as strong among planners as among draughtsmen and laboratory staff. The returns from

Table 6.10 Percentage of technical staff considering the shop floor differential insufficient, by occupational group, level of qualification and level of salary

Salary level	Differential insufficient	Differential[1] sufficient	Level of qualification	Differential insufficient	Differential[1] sufficient
1. Draughtsmen					
	£				
(a) 1,600 and over	95	0	Degree	100	0
1,500–1,599	91	0	H.N.D.	100	0
(b) 1,400–1,499	83	8	H.N.C.	91	13
1,300–1,399	100	0	O.N.D.	0	0
1,200–1,299	83	4	O.N.C.	97	3
(c) 1,100–1,199	75	0	City and Guilds	71	14
1,000–1,099	70	30	Final City and Guilds	100	0
(d) 900– 999	100	0	Other City and Guilds	0	0
Less than 900	100	0	No qualification	96	4
2. Laboratory Staff					
	£				
(a) 1,600 and over	94	6	Degree	50	1
1,500–1,599	80	20	H.N.D.	50	20
(b) 1,400–1,499	100	0	H.N.C.	86	8
1,300–1,399	90	0	O.N.D.	0	1
1,200–1,299	100	0	O.N.C.	83	12
(c) 1,100–1,199	80	12	City and Guilds	75	25
1,000–1,099	88	8	Final City and Guilds	80	0
(d) 900– 999	90	0	Other City and Guilds	100	0
Less than 900	80	14	No qualification	90	10
3. Planning and Production					
	£				
(a) 1,600 and over	67	17	Degree	50	50
1,500–1,599	50	50	H.N.D.	100	0
(b) 1,400–1,499	60	14	H.N.C.	90	10
1,300–1,399	82	18	O.N.D.	0	0
1,200–1,299	67	33	O.N.C.	80	10
(c) 1,100–1,199	86	14	City and Guilds Final	0	0
1,000–1,099	25	75	City and Guilds	63	37
(d) 900– 999	66	34	Other City and Guilds	60	100
Less than 900	100	0	No qualification	64	36
4. Quality-Control Technicians					
	£				
(a) 1,600 and over	75	12	Degree	100	0
1,500–1,599	100	0	H.N.D.	100	0
(b) 1,400–1,499	100	0	H.N.C.	100	0
1,300–1,399	100	0	O.N.D.	100	0
1,200–1,299	67	17	O.N.C.	60	14
(c) 1,100–1,199	83	0	City and Guilds Final	92	0
1,000–1,099	70	10	City and Guilds Full	34	66
(d) 900– 999	70	20	Other City and Guilds	100	0
Less than 900	100	0	No qualification	80	9

(1) Total of 100 made up by *not sure*.　　　　　SOURCE: Questionnaire Survey

those earning between £1,000–£1,099 showed more satisfaction with the differential than dissatisfaction. Significance tests showed differences at all levels of salary not to be a critical factor in determining attitudes to the shop-floor differential. When attitudes of planners were examined in the light of levels of qualification, there was evidence that dissatisfaction was greater among the more highly qualified than among the less qualified.

The replies from quality-control technicians showed a greater dissatisfaction with the differential among those earning between £1,300–£1,599 than those earning between £900–£1,299. Attitudes in relation to educational qualifications showed the greatest dissatisfaction among the relatively mostly highly qualified.

The conclusion which emerges is that for all groups the level of salary was not significant in explaining different attitudes to the shop-floor differential. The dissatisfaction of those at the top of the salary range probably arose from their perception of the recent growth of manual workers' earnings. There was a feeling that these workers had been able to use the 'productivity' exception to the Incomes Policy norm, by selling restrictive practices, to get large increases. Higher salary groups considered that they had no restrictive practices to sell and therefore in the past five years had found it difficult to obtain salary increases above the rate of increase laid down by Incomes Policy White Papers. An explanation of the discontent of those at the bottom of the salary scale probably lay in a feeling that craft workers with overtime payments, piece-rate payments, etc., were able to increase their earnings more than young technicians who had to spend leisure time, not on overtime working but on studying to pass examinations to move up a company's hierarchy of skills.

An analysis of the overall relationship between qualifications and attitudes to salary differentials showed that technicians with H.N.D. and H.N.C. except for laboratory staff, were strong in their condemnation of the narrowness of the differential from the shop-floor. This was possibly explained by a feeling that years of study to achieve greater skills for the company's benefit were not properly rewarded, since the non-payment of overtime or bonuses to technicians (or the payment of reduced rates in relation to those of manual workers) enabled the manual worker to obtain a high percentage of the technician rate. Except among planning and production engineers, there was a strong dissatisfaction with the shop-floor differential among those having no formal qualification. This attitude perhaps stemmed from a feeling that to join the white-collar staff of a company was considered by society as a surface sign of improving one's lot in life. Many of these less-qualified technicians came from manual jobs, believing that they would be earning more money by doing this. It was natural, therefore, for them to consider themselves to be more beneficial to the company than the shop-floor workers and expect to see this reflected in pay differences.

Market and institutional influences on salary structure

Technicians had relatively little knowledge of the facts of their labour market. At two engineering plants in Lancashire there was an ignorance of the level of salaries for similar occupations in other parts of the country. Technicians had a vague feeling that salaries were perhaps higher in some areas, but dismissed the difference as being necessary to compensate for the higher cost of living in these regions. The views of these technicians were based on impressions, since they had no detailed knowledge of whether the cost of living was in fact higher, and if so whether the increased salaries more than compensated for the higher prices. The situation could reflect the lack of trade unionism among technicians. There were no bodies like Joint Shop Steward Combine Committees to keep rates of pay at different companies for similar occupations under review and then to use collective bargaining machinery to achieve parity of rates between companies and regions. There are signs that the unions seeking to organize technicians are beginning to make efforts to overcome this problem of labour-market information. The A.S.T.M.S. is making inter-plant comparisons of salaries of its members employed in some big companies. The draughtsmen's union journal publishes wage settlements made by its members in different parts of the country. For some draughtsmen the imper-fection of the market through lack of knowledge has been reduced by becoming contract draughtsmen, whereby the agency with which they register searches for the highest-paying jobs and passes the information to the registered draughtsman. Whether the long-run effect of a greater flow of information will lead to a general levelling of technicians' salaries in different regions depends on institutional as well as pure economic factors.

Labour economists have long been interested in the question of whether wage determination is the result of power forces or market forces. There are obvious methodological and statistical difficulties involved in this question, but the research in this study did appear to illustrate the influence of both economic and

G

institutional forces. Most technical staff was not unionized, but some of the broad occupational groups did exhibit a relatively high degree of unionization, for example, draughtsmen 43 per cent and laboratory staffs 56 per cent. An examination of the spread of salaries between relatively unionized and non-union-ized groups might give a guide to the possible effects of trade unionism. It could be hypothesized that the spread of earnings would be least among the highly unionized groups because of the union policy of demanding the rate for the job and not the person.

Table 6.11 shows the degree of unionization among tech-nicians and the spread of earnings for four of the occupational groups. The expected relationship of the smallest spread of earn-ings among the most highly unionized did not appear. From the Table the relationship if anything appeared to be inverse. The smallest spread of earnings was found in a sub-group of the sample which happened to be the least unionized, and the largest among the most highly organized, namely the laboratory staffs.

However, it did not follow that trade unionism had no effect on technicians' salaries, and the field researches revealed situa-tions in which the presence of trade unions had some influence on management attitudes. At four of the plants (two in vehicles and two in engineering) a number of line managers argued that there was no need for their personnel to join a trade union, because it was company policy to pay rates of pay above the minimum negotiated with the technicians' unions, in order to make trade-union membership unattractive. One manager said: 'The office was strongly organized in the early 1960s. Some new members of the drawing office came to the plant as DATA members, but now seem to drop out quite quickly. We pay above DATA rates and there are good conditions, so people have nothing to gain from union membership.'

Another manager in the London area suggested that union membership was unnecessary because the tightness of the labour market meant it was possible to get wage increases without the support of collective bargaining. Again, a line manager remarked: 'There is no profit for technicians in belonging to DATA or any other union. Our company pays above the flat rate negotiated by DATA with the E.E.F. and there is no real need to join.'

These managers were arguing that union membership was

unnecessary because pay was above the union rates. However, the technicians' unions may have had an influence on earnings in that management was prepared to pay wages above the going rate in order to keep the union out of its plants. Some employers were ready to do this when they considered the cost of paying above the union rate to be less than accepting unions and having to establish collective bargaining machinery.

Table 6.11   Degree of unionization among technicians, and the spread of earnings

Occupational Group	Percentage union members	Percentage non-union members	Mean quartile deviation of earnings (£)	Coefficient of quartile deviation
Laboratory staff	56	41	249	.21
Draughtsmen	43	51	194	.14
Planning and production engineers	17	62	185	.14
Quality-control technicians	37	57	202	.17

SOURCE: Questionnaire Survey.

Militant action by technicians' unions appeared to have had an influence in some cases. There were examples of a technicians' trade union being able to force wages above the level the employer had sought to pay. Technical assistants employed by an aircraft company joined DATA and approached management for the payment of DATA minimum rates. When resistance was met with the union called a one-day strike in support of the assistants, which was relatively successful. One firm in metal manufacture had a virtually monopsonistic position in the labour market in the region. Its only competitor for labour was a recently established light engineering firm, which provided few employment opportunities for draughtsmen and technicians. Being in this strong position it was possible for the company to pay lower rates than those paid at other firms in the North-East. There was a feeling among technicians that the company was taking advantage of its bargaining position, though they were unable to demonstrate it. A comparison of earnings with those of draughtsmen and technicians employed in other metal companies in the North-East in fact showed the company in an unfavourable light. One draughtsman stated clearly: 'This company is perhaps paying less than it ought because of its isolated geographical position and the nature of the local labour market.' Draughtsmen at the company were members of DATA, and before the union's 1965 campaign to get a national minimum wage for age, they were receiving £2–£3 below the basic rates the union was attempting to establish. In 1965 DATA draughtsmen withdrew their labour in an attempt to force the company to pay the union minimum rate. The strike lasted for five weeks, but it achieved the purpose of forcing the company to pay DATA minimum rates.

A further example of trade unionism influencing salary levels was found in the South-West region of the country. The firm had its group headquarters in London. There was liaison between 'stewards' in the plant and those in company headquarters belonging to the same staff union. Claims for increases in salary or changes in working conditions were submitted in all union-organized plants in the group simultaneously. A settlement at headquarters in London (an area of excess demand for technicians) preceded other settlements and indicated the acceptable level of change elsewhere. Before arriving at a final settlement at this company in the South-West there were numerous local issues which could be included as factors influencing the way in which the headquarters settlement was accepted locally. Union tactics in local negotiations were largely determined by a knowledge of the course of negotiations at the group headquarters. The personnel manager confessed that inter-plant communications were better among staff-union representatives than among management.

The influence of incomes policy

Trade unions are not the only institutions that may influence wage levels or the rate of increase in money incomes. After 1948 there was a growing interference by Government in wage matters through influence and persuasion: the holding down of wages by voluntary agreement. When the pay-pause policy collapsed other steps were taken by the establishment, first, of the National Incomes Commission and later of the National Board for Prices and Incomes. After April 1965 the aim was to pursue a comprehensive policy controlling the rate of growth of prices, productivity and incomes through the National Board for Prices and Incomes.

The technicians' unions made their views on the incomes policy clear in early 1965 in a pamphlet entitled 'A Declaration of Dissent'. This document argued that Britain's economic problems were the result, not of wage inflation, but of the misdirection of economic resources, caused by the pull of market forces within the economy frustrating basic social needs, and by too heavy a burden of military expenditure. The pamphlet argued that '... the British system of rewards needs an overhaul. It provides little incentive where incentive is most needed, and contributes to the gross inequalities which disfigure our society...'[1] The unions did not accept the 3½ per cent norm of the first phase of the Productivity, Prices and Incomes Policy as being adequate for their members, and argued that their function was to secure for their members adequate pay for their training, skill and experience, since they considered that in Britain wages and salaries were too low, skills under-rewarded, and qualifications inadequately remunerated.

In the engineering industry the incomes policy did impinge directly on technicians. In August 1965 DATA agreed with the Engineering Employers' Federation a scale of minimum rates for draughtsmen, but in June 1966 a claim was submitted for a new national agreement. The employers suggested that in the light of the Government's incomes policy the claim should remain on the table, to be taken up again in twelve months' time. Negotiations continued but no agreement was forthcoming, and DATA gave notice to the Federation of termination of the national agreement from September 1967. Claims were pursued on a domestic basis and a number of these settlements were referred to the National Board for Prices and Incomes, which reported that it considered them to be a breach of policy. Eventually a new agreement was signed in the summer of 1968.

The E.E.F. had made an agreement with the A.Sc.W. in 1965, and the incomes policy delayed the negotiation of a new agreement. In December 1967 the National Board for Prices and Incomes reported on the negotiations between the E.E.F. and the technicians' unions.[2] The report recommended that immediate steps should be taken towards the development of proper

---

[1] *A Declaration of Dissent: Technicians' Union and Incomes Policy* (1965), p. 1.
[2] National Board for Prices and Incomes, Report No. 49, *Pay and Conditions of Service of Engineering Workers* (1st Report), Cmnd 3495 (H.M.S.O., London, December 1967).

salary structures for technicians. At national level the Board
considered there should be an agreement on definitions of
categories of staff workers, and on the basis of grading schemes
through joint discussions between the E.E.F. and the staff
unions. Eventually there should be a negotiation of realistic
minimum salary scales. The dislike of technicians for the Productivity, Prices and Incomes Policy came over strongly in this
research project. The majority of technicians saw the policy
as holding down wages and salaries. Only in one plant, where
a productivity bargain was being negotiated between the management and manual workers, and management and technicians, did
the productivity aspect of the policy seem to be known. Technicians in an engineering plant in Lancashire considered
the policy was discriminating against them in relation to the
manual workers. The comment of laboratory technician typifies
this:

> ... the Prices and Incomes policy has had an impact on discontent
> and is probably a major factor in the growth of staff unionism in
> the plant. There is a feeling that the firm has tended to hide behind
> the policy. The shop floor workers have found ways through this
> and as a result differentials between the two groups have
> narrowed. ...

In the same plant the DATA 'corresponding' member considered
that the incomes policy would give his union an excellent opportunity to increase its membership in the firm. The policy had
meant that annual salary reviews had become subject to the
norm set down by the Government. To exploit this unrest the
union needed some members to work among the unorganized
section of draughtsmen, and he was optimistic concerning the
results.

On the other hand, a manager of an electronics firm in London
considered that the incomes policy was having little effect
because of the imbalance of the labour market in the area. There
were about ten similar electronics companies in the area and
competition for labour between these firms was intense. The
prices and incomes policy was being flouted daily by technicians
taking advantage of the tight labour market to change companies
for a similar job at a higher salary.

The strongest anti-incomes policy comments were made at two
oil refineries of the same company, but in different areas of the

country. In both plants the management was attempting to negotiate productivity bargains with craft workers, general workers and technicians. In June 1968 the company attempted to introduce a productivity deal for technical staff. As a result of the deal a 5 per cent increase in salary was awarded, back-dated to 1st April 1968. This was above the norm of the phase of the incomes policy at that time, but the agreement was accepted by the Department of Employment because of past flexibility on the part of technicians, and the productivity elements in the present deal. These included job flexibility (within and between departments), time flexibility (whereby the normal numbers of working hours may be worked at varying times of the day or week) and a disturbance allowance for shift-working. At the same time the company entered into negotiations with its craft workers whereby in return for productivity concessions, much larger pay increases were granted. A third set of negotiations with the T.G.W.U. members and the ensuing settlement also resulted in a pay award above that of the technical staff. A fourth set of negotiations was undertaken with laboratory assistants, who were offered 5 per cent in return for productivity concessions.

These proposed deals brought strong reactions from technicians. A number of laboratory assistants complained that they were being unfairly treated. They considered the policy was penalizing them for their flexibility in the past, so that now they had no 'restrictive practices' to sell in return for higher wages. The craft and general workers were seen to be gaining from their former stubbornness. One inspector, however, argued that he was not alarmed by this trend because the elimination of restrictive practices was a thing to be welcomed whatever the price. Asked if he thought membership of a trade union would be a good way of raising technicians' salaries, he answered that he would never consider joining any trade union because they *all* had too many restrictive practices. This attitude might have been fostered by the inspector's previous job. He had been an engineer on a sea-going vessel and his contact with unions in the past had been in British ports and shipyards, areas where trade-union restrictive practices are very strong.

A further complaint was that the policy discriminated against technicians because in their jobs productivity measurement was difficult. Draughtsmen argued that their job was ancillary to production, and that it was thus difficult to measure their direct contribution to the productivity of a firm. Work study has been advocated as a means of improving drawing-office output, but DATA opposed this management technique. In its evidence to the Royal Commission on Trade Unions and Employers' Associations the union explained its case.

> ... This is not because it [the union] is opposed to efficiency, but because it does not believe that techniques of this kind are suitable for application to work where quality is of supreme importance. A draughtsman who thinks at length about a project, or who goes into the workshop to seek advice, may be making a much greater contribution than a draughtsman who fills his board with a great many lines for a very inferior design ...[1]

The union, however, is not opposed to discussions on efficiency, for example on ways of improving the allocation of work, the filing system for drawings, and the location of the drawing files and print room. The National Board for Prices and Incomes Report No. 68[2] suggested a number of ways in which the productivity of drawing offices could be improved by the more effective use of manpower. The Board pointed to the relationship between the drawing office and other activities in the company and suggested that productivity could be improved by integrating the drawing office more effectively into the entire system of work planning and control, for example the information flow in and out of the office and lines of communication between departments. It argued that improvements could be made by reorganizing work within the drawing office, for example delegating routine clerical duties to less skilled manpower; basing standards on the newer techniques available, rather than wholly on past experience, thus reducing the cost per drawing; and introducing more simplified or fractional draughting rather than adhering to traditional practices. The DATA journal, in commenting on the report, perhaps significantly made no comment on these suggestions.

Both N.B.P.I. reports threw light on a problem area for technicians, namely the lack of national salary structures, and suggested that there should be a proper development of salary

---

[1] Royal Commission on Trade Unions and Employers' Associations, *Minutes of Evidence* 36 DATA (H.M.S.O., London 1966).

[2] National Board for Prices and Incomes, Report No. 68, *Agreement Made between Certain Engineering Firms and DATA* Cmnd. 3632 (H.M.S.O., London, May 1968), paras. 34–40 and Appendix E.

structures for technicians based on grading systems that give recognition to skill and responsibility. It is perhaps ironic that a policy which was attacked by the technicians' unions because it would continue the lack of rewards for skill and responsibility should become the ally of the unions on this point.

In an attempt to find out how technicicians considered market and institutional forces influenced their pay, a question was inserted into the questionnaire asking them to pick three factors from a list of nine, indicating what they felt were the main reasons why they received salary increases. Of these nine factors, four were institutional (individual bargaining, collective bargaining, changes in the wage structure and incomes policy), three economic (the demand for technicians in other companies, management deciding what a technician is worth in relation to skills, etc., and the financial position of the firm [that is, its ability to pay]), and two miscellaneous.

The opinions of 70 per cent of the sample who answered the questionnaire completely were divided almost equally between institutional and economic factors: thus 34 per cent considered that institutional forces were the main reasons why they received salary increases (13 per cent ascribing them to the fact that the shop floor had already had an increase), and 44 per cent felt that economic factors were the main reasons (13 per cent ascribing them to management's evaluation of their worth).

On an occupational basis, 40 per cent of laboratory staffs considered institutional factors to be the reason for their salary increases. Twenty-seven per cent thought market forces important, the demand for technicians in other companies coming out highest. Forty per cent of the replies from draughtsmen gave institutional reasons for obtaining increases, with 15 per cent ascribing them to the bargaining skill of DATA, and this is perhaps a reason why 43 per cent of draughtsmen in our sample were members of this union. Thirty-three per cent considered market forces the main reasons.

The returns from planning and production engineers, indicated that 36 per cent thought institutional factors led to their salary increases, and 43 per cent market forces. The replies from quality-control technicians were equally divided between institutional (34 per cent) and market forces (33 per cent); 11 per cent thought individual bargaining came first, 10 per cent trade-union bargaining skill, and 13 per cent, increases on the shop floor. Eleven per cent thought the demand for technicians in other companies was important, 8 per cent the financial position of the firm, and 13 per cent the decision of management as to what the technician was worth. Technicians therefore appear to see both institutional and market forces as determinants of their salary increases, but for three of the main four groups, institutional forces appear to be slightly more important.

Fringe benefits

The remunerative conditions of service have been fully discussed, but conditions of employment such as the length of the working week, the amount of annual paid holiday, pension schemes, sick payment and security of employment have been left unmentioned. These latter matters are often referred to as 'fringe benefits', but the questionnaire response to fringe-benefit matters was poor. Interviews revealed varying impressions as to what constituted a fringe benefit. Some saw it as a 'perk' whereby one could obtain money from the firm through expense accounts, or receive a payment on which income tax could be avoided. These results would appear to bear out the observation of G. L. Reid and D. J. Robertson in *Fringe Benefits, Labour Costs and Social Security*: '. . . the term "fringe benefits" has a somewhat puzzling and fanciful aspect to the uninitiated and even to those who use it frequently it is elusive and diverse in its meanings . . .'[1] There are no generally recognized definitions of a fringe benefit. Economists are interested in such benefits because they add to the costs of the employer and may affect his output–pricing policy, and demand for labour. Fringe benefits may be described as items which add to an employer's labour costs and provide a benefit, usually at no cost, to the employee.[2] This is a wide definition, and would include provisions like separate refreshment facilities for different groups of workers.

The questionnaire results and visits to the plants participat-

[1] G. L. Reid and D. J. Robertson (eds.), *Fringe Benefits, Labour Costs and Social Security* (Allen and Unwin, London 1964).

[2] This is not strictly true of pension schemes since very few companies have non-contributory pension schemes and employees are generally asked to make some contribution towards their retirement pension.

ing in the study gave some indication of the nature and extent
of fringe benefits enjoyed by technicians. In the chemical
industry the length of the working week for technicians was
37½ hours, compared with 40 hours for manual workers. Technical staff received three weeks' annual paid holiday, increasing
to four after ten years' service. Manual workers on the other
hand received only three weeks' paid holiday regardless of length
of service. Technicians received sick pay and were members of
a company pension scheme, but such schemes were common
to all grades of workers.

In metal manufacture one company provided a pension scheme
for both staff and manual workers. The scheme for staff was
contributory, with each staff member paying 4½ per cent of his
annual gross salary less £250, on top of which the company
made a contribution. The maximum pension on retirement was
two-thirds pay. The amount of paid annual holiday leave varied
with the length of service. For those in the company's employ
for up to five years, the amount was two weeks and three days;
those with up to ten years' service received three weeks and
three days, while technicians with more than ten years' service
received four weeks. There was also a sickness scheme whereby
the amount of pay during periods of sickness depended upon
the length of service. A long-serving member of the technical
staff could have almost twelve months off with full pay before
dropping back on to half pay.

Firms in vehicles components and engineering were in the
same collective bargaining unit, but not all the companies in
the study were federated to the E.E.F. DATA conducted
'guerilla' tactics against federated employers in an attempt to
force them to concede four weeks' annual paid holiday. The
success of the campaign led the E.E.F. to propose a charter of
standard conditions of employment for staff workers in August
1966. The federation proposed that the charter should apply to
all staff workers, whether full-time or part-time, except men
earning over £1,500 and women earning over £1,200. The charter
proposed a 37½-hour week and fifteen days' annual holiday plus
six bank holidays, payment for redundancy to be made in accord-
ance with the Redundancy Payments Act 1965, changes in the
length of time required to terminate employment, sickness payments related to length of service (giving the maximum payments in any twelve months), and suggestions on pay for over-

time and shift working. The charter stressed that these were minimum conditions, and that in firms where more favourable conditions existed they should remain in force. The technicians' unions rejected the proposals because they were not prepared to have their members shackled to the pace of the slower and more inefficient firms, and to have the conditions of their members in more prosperous and often technically more advanced firms held back.

The N.B.P.I. undertook a survey of the conditions of service of staff workers in the engineering industry.[1] It found that 72 per cent of technicians worked under 40 hours, and 60 per cent of all staff workers (including clerks) worked in firms with a basic standard of fifteen days' or more annual holiday. One-third of the companies provided notice of termination of employment in excess of the Contract of Employment Act 1963, and 6 per cent of firms made payments in excess of the requirements of the Redundancy Payments Act 1965.

The questionnaire sought to secure information on which fringe benefit technicians considered to be the most beneficial to them. Forty-one per cent of the respondents did not answer the question, and the most highly prized fringe benefits were in the miscellaneous category 'other', followed by pension schemes, sickness payments, and holidays. Nearly 13 per cent answered that membership of a pension scheme was their most prized fringe benefit, and the same percentage considered that it was a sickness payment scheme operated by their company. Nine per cent replied that theirs was holidays with pay. When the question is analysed on an occupational basis the same trend appears. If the miscellaneous category of 'other' is excluded, then for the four main occupational groups for which it was possible to obtain data, sickness and pension schemes are seen as the most prized fringe benefits. Planning and production engineers were the exception. Amongst this group the most prized fringe benefit was holidays, whilst membership of a pension scheme came out below the benefit of more freedom to take time off among this group.

It is difficult to explain attitudes to fringe benefits, because the value attached to such benefits obviously varies from indi-

[1] National Board for Prices and Incomes, Report No. 49, *Pay and Conditions of Service of Engineering Workers* (1st Report), Cmd. 3495, and the *Statistical Supplement*, Cmnd. 3495-1 (H.M.S.O., London 1967).

vidual to individual. There are periods when individuals lose their ability to earn income, such as in sickness and old age, and it is natural that technicians should prize schemes that provide incomes during such periods, rather than those that do not, for example more holidays and a shorter working week. In a period of unemployment when individuals lose their ability to earn, but technicians probably see their staff jobs as being more secure than those of manual workers, and this may account for 'more security' receiving a poor response in the survey. In other words, they may not see unemployment as an urgent prospect for themselves and therefore do not value security of employment in the same light as security during sickness and old age.

The relative importance of fringe benefits to technicians may be influenced by a number of factors, for example the level of salary. A hypothesis tested was that the higher the level of salary the less importance attached to fringe benefits. For those earning above £1,400 the importance of fringe benefits did appear to fall with increasing salary, but at salaries below this the pattern was haphazard. The figures for laboratory staff also showed that the expected simple relationship between the importance of fringe benefits and the level of salary did not hold. Except for laboratory staff earning between £1,500 and £1,599, fringe benefits appeared to have a low importance amongst laboratory technicians. In the salary band £1,200–£1,299, 62 per cent answered that fringe benefits were either quite unimportant or very unimportant to them. The returns from planning and production-control engineers had an uneven pattern of attitudes to fringe benefits by level of salary. From the results of the survey it would appear that there is little relationship between importance of fringe benefits as they are perceived by technicians and the level of technicians' salaries. The attitudes of quality-control technicians to fringe benefits, in relation to the level of salary, also showed no simple trend. The importance attached to fringe benefits appeared to be much higher than amongst laboratory staffs.

Trade-union membership could be a factor influencing attitudes to fringe benefits, since technicians' unions provide friendly-society benefits for their members. It was possible that this might have led trade unionists to consider the need for fringe benefits to be less than amongst technicians who are non-unionists; however, the returns showed no clear pattern and varied from group to group.

Amongst draughtsmen, DATA members did appear to consider fringe benefits less important than non-DATA members. Among laboratory technicians (the most highly unionized group) the trend appeared to be the reverse. Trade-union members appeared to favour fringe benefits more than non-unionists. Fifty per cent of those who were members of unions considered benefits to be very important or quite important, but only 44 per cent of non-unionists. Only 25 per cent of trade unionists thought fringe benefits to be either quite unimportant or very unimportant, as opposed to 33⅓ per cent of non-union laboratory technicians. Quality-control technicians who were members of trade unions considered fringe benefits more important than those who were not members. Eighty per cent of unionists replied that fringe benefits were either very important or quite important to them and only 14 per cent considered them unimportant. Of the non-unionists, however, only 57 per cent thought fringe benefits important, while 30 per cent replied that they were either quite unimportant or very unimportant. On the whole it would appear that if a technician is a member of a trade union he is less likely to consider fringe benefits to be important than if he is a non-union member. However, this is a tentative conclusion, and investigation with a greater sample response would be necessary before any firmer statement could be made.

The traditional view of fringe benefits has been to see them as a reward to staff members of a company. By 'staff' was implied those who had stepped from the shop-floor grades into white-collar categories having a direct responsibility and relationship to management. For close associates of management, with a greater commitment to the company, the firm was prepared to provide rewards in the form of a pension scheme, holidays without loss of pay, and sickness schemes, etc. In the past, management has been able to explain away the narrowing of wage differentials between staff and manual workers by pointing to the differences in relative fringe benefits, which are also considered to be representative of status within the company's hierarchy. It was common to find the trend among the plants involved in this research was towards uniformity of fringe benefits between manual workers and staff workers.[1] The attitude of technicians to this trend appeared from the questionnaire replies to be favourable, perhaps because they did not realize the implications in terms of status.

A number of those interviewed did make various complaints. One technician said it was not seen as promotion in terms of fringe benefits to move from a shop-floor position to become a technician, whilst another one argued that uniform fringe benefits would reduce the incentive to try and get into management rather than languish on the shop floor. Fifty-seven per cent of technical staff were either 'very much' or 'quite' in favour of a policy of extending to manual workers the fringe benefits generally enjoyed by employees of weekly-staff status. Only 23 per cent said that they were not really in favour, or not at all in favour. Fifty-eight per cent of the laboratory returns showed favour towards the extension of fringe benefits, and only 21 per cent expressed disfavour. Among service engineers, 38 per cent answered that they were either not really in favour or not at all in favour of the extension of fringe benefits to workers of weekly-staff status. Thirty-three per cent of supervisory grades and 30 per cent of planning and production engineers expressed opposition to the extension of benefits to workers thought to have lower status.

Thus an analysis of technicians' attitudes to fringe benefits showed that while greater importance was attached to those that provided income at periods of income loss, such as old age or sickness, rather than holidays with pay and freedom to take time off, benefits in general were regarded by technicians as less important than salary levels.

Conclusions

If the trend towards uniform non-monetary benefits continues, the question will arise as to how status will be accorded to various groups by management. If there are common benefits between workers, then wage relativities and promotion opportunities may become the measures of status. Salaries will be seen as indicating management evaluation of technicians in relation

[1] This finding also emerges from an N.B.P.I. survey: National Board for Prices and Incomes, Report No. 49, *Pay and Conditions of Service of Engineering Workers*, Cmd 3495, and the *Statistical Supplement*, Cmd 3495-1 (H.M.S.O., London 1967).

to other groups in the plant. Managements may find themselves
faced not only with 'leap-frogging' claims from manual workers,
but also claims from staff workers for the restoration of differentials that reflect skill and responsibility. These claims may be
extended as the result of the increasing degree of unionization
among technicians. If this trend also continues, possibly helped
by Government policy,[1] so that collective bargaining develops,
then negotiations for 'proper differentials' between technicians
and manual workers may become an everyday thing.[2]

Management's problem will be to reform its existing wage
structure on the basis of job evaluation so that any new structure
is based on an appraisal of skills, and the requirements of the
job. One plant participating in the study had realized this and
had evaluated its technicians' and technologists' jobs, and was in
the process of evaluating shop-floor jobs. It was anxious, however, about the outcome since preliminary results suggested
that skilled manual workers could come out above the lower-
grade technicians. This one example illustrates the magnitude
of the problem.

The questions of technicians' salaries and conditions of
employment are seen to be closely related to the questions of
technicians' low status that were discussed in a previous section.
It has been shown that the traditional means of demonstrating
status within a plant, through staff benefits and privileges, have
been becoming less significant. When single-status plants are
established, it is likely that wage–salary relativities will become
the dominant indicators of remaining status differences.

---

[1] The Government's Industrial Relations Act of 1971 gives a statutory right
to join a registered trade union.
[2] A.S.T.M.S. has already stated that it wishes to establish a proper differential
between salaried staffs and manual workers.

7 Technicians and Mobility

When the work goals which technicians desire to achieve are blocked, they may seek to overcome the obstacles they face in a number of different ways. They may decide to move to another job which they hope to find more advantageous. Alternatively, they may try to bring about an improvement in their employment situation by qualifying for promotion to a higher grade or greater responsibility, or by joining a trade union or professional association. On the other hand they may just apathetically accept the situation.

Factors internal to the plant, for example, the availability of career development, the wage structure, the degree of satisfaction from work, and management policy towards technicians, may give rise to a situation in which a decision to move to another job becomes a considered course of action. Discontent with the work situation does not automatically mean movement, and if a technician does consider this course, factors external to the work-place may prevent mobility. Such factors can be classified under the headings 'economic' and 'social'. The most important economic constraint is the state of the labour market for technicians. The availability of jobs depends upon the general level of economic activity and the extent of the labour market (local, regional, national and international) for the particular occupation. The individual's perception of the labour market is important, and influenced by a number of factors, but particularly the technician's knowledge of the market and his strength of conviction that better opportunities are available. These factors limit the size of the labour market, because costs are involved in moving and the individual will expect to recoup at least these costs. The community within which the individual lives provides a social restraint, since the benefits from participating in the community may outweigh the disadvantages of the work situation.[1]

In this chapter we shall examine the extent to which the work situation, through the failure to fulfil the expectation of technicians, gives rise to potential mobility, and the extent to which mobility has taken place.

The concept of labour mobility

The economist is concerned with mobility from the point of view of the most efficient allocation of economic resources. Economic theory predicts that labour will move from low-wage areas to high-wage areas. Economists have qualified this view to argue that the nature of occupations will attract some people but deter others, and that what is equalized is not the wage rate but the net advantages of occupations. Few economists, however, would claim that labour mobility is simply a reaction to higher financial rewards in other occupations,[2] and empirical work by W. B. Reddaway[3] has shown that changes in the demand for labour in various industries and occupations operate to secure a redistribution of the labour force mainly through direct changes in the 'job opportunities' made available rather than wages differentials. A theory of labour mobility as treated by economists is incomplete in that it ignores the influence of sociological factors, which may be strong enough to obstruct any movement in response to economic incentives. However, this does not mean money is unimportant in causing movement from one work situation to another, and research on 'non-economic' factors in the motivation of workers has not found financial rewards unimportant. A theory of labour mobility, therefore, must comprehend both the economic and the sociological aspects of job-changing.[4]

[1] This can take many forms, varying from political activities such as membership of the town council to the tie of personal friendship built up over a long period of years.

[2] See Lloyd G. Reynolds, *The Structure of Labor Markets* (Harper, New York 1951).

[3] See W. B. Reddaway, 'Wage Flexibility and the Distribution of Labour', *Lloyd's Bank Review*, 1959, pp. 32–48. Also H. S. Parnes, *Research on Labour Mobility* (Social Science Research Council, New York 1954).

[4] J. H. Smith, 'The Analysis of Labour Mobility', in B. C. Roberts and J. H. Smith (eds.), *Manpower Policy and Employment Trends* (L.S.E. – Bell, London 1966).

It is possible to distinguish three aspects of labour mobility. There is the capacity or ability of workers to move from one job to another. This will depend upon whether an individual possesses general or specific skills, but from an operational standpoint there are difficulties in giving meaning to this concept.[1] Secondly, there is a worker's willingness or propensity to move given the opportunity. A worker's ability to make a given job change is no assurance that he will choose to make it, as he may be unwilling to make the change. Mobility viewed as the propensity to move also gives rise to measurement problems. Finally, there is mobility as ascertained by an historical examination of past movements, and such a method is adopted in this study.

Labour mobility can take many forms. It may be a movement into and out of the labour force, movement from employment to unemployment, a change of department, a change of grade, a change of occupation, or a change of industry. Any one actual movement may involve a combination of any of these. In the analysis of mobility among technicians concern will be with movement within an establishment or firm (i.e. the internal labour market) as well as between employers.

The concept of a labour market

The popular view of a market is a place where buyers and sellers meet for exchange. A 'market' in economics is an abstract concept, and the labour market refers merely to the broad area within which exchange between buyers and sellers takes place. The areas vary, depending upon the nature of work and the mobility of labour, but it is possible to distinguish four broad types of

---

[1] The ability to move gives some guide to potential mobility. L. Broom and J. H. Smith, 'Bridging Occupations', *British Journal of Sociology*, December 1963, attempting to identify the potentiality for mobility, introduced the concept of an occupational system. The essence of this is that skills acquired in one job may govern further job-choice and mobility. Six types of occupation systems were identified – bridging, closed, preparatory, career hierarchical, incremental hierarchical, and residual. Technicians' jobs are 'bridging' in the sense that they are a stepping stone to technologists' jobs. Some technicians' jobs (e.g. those of testers and inspectors) are closed occupations, while some others (e.g. those of young laboratory assistants) are preparatory. Draughting and industrial engineering are the hierarchical forms. In this respect the concept would be difficult to apply to technicians.

external labour market, local, regional, national and international, which are overlapping and interdependent.[1]

There is also the internal labour market, which refers to the labour available to the firms within the establishment. After the point of entry into the firm management can adjust labour supplies by up-grading or down-grading employees already on the payroll. If there are not sufficient supplies within the internal market then the company must recruit from the external market. Technicians' occupations can be viewed on a continuum, ranging from laboratory assistants, executing routine tests, at one end, to designers with high academic qualifications and responsibility at the other. The lower-level technicians employed in routine laboratory work surveyed in this study had been recruited mainly from the local labour market. For example, 40 per cent of the laboratory staffs had joined their present firms straight from school. On the other hand the medium grade of technicians (for example, testers and inspectors) appeared to have been mainly drawn from a regional market. Less than 10 per cent in the sample had reached their present position by vertical promotion within their firm. High-level technicians engaged in specialized and complex work were recruited from the national and sometimes even the international labour market. The international labour market was of significance to draughtsmen and technicians in aircraft manufacture, many of whom were recruited by American aerospace firms.[2] Draughtsmen's vacancies and designing and work-study jobs tend to be advertised on a national scale, as the Situations Vacant sections of the daily and Sunday newspapers amply demonstrate. However, the international market is more appropriate to technologists than technicians.

The technicians' labour market covers a wide area, and the question arises as to the boundaries between, at the upper end, the technicians' and technologists' labour market, and, at the

---

[1] Attempts have recently been made to determine the size of the local labour market on the basis of a travel to work area. See, for example, J. F. B. Goodman, 'The Definition and Analysis of Local Labour Markets: Some Empirical Problems', *British Journal of Industrial Relations*, Vol. VIII, No. 2 (July 1970), and Derek Robinson, *Local Labour Markets and Wage Structures*, (Gower Press, London 1970).

[2] The Society of British Aerospace Companies, in a report on the 'brain drain' during 1966, estimated that from a survey of 88 companies, 34 per cent of draughtsmen went overseas or joined foreign-owned firms in the U.K. Over 50 per cent of the 34 per cent went to the United States.

lower end, the technicians' and manual workers' labour market.[1] There is a considerable blurring of occupational differences at the boundaries, and the case studies threw up examples of technicians doing technologists' work but receiving technicians' levels of pay. In a number of firms it was deliberate policy to allow only employees with graduate qualifications to move into the technologists' labour market. The practice of employing technicians as technologists appeared to vary with the relative supply and demand for technicians and technologists. In the North-East, firms complained of difficulty in attracting graduates, and the general scarcity of jobs in the area enabled companies to use technicians as technologists. In the London and South-East region, however, the supply and demand for graduates was greater and the practice of using technicians as technologists was rare. Although technological developments had widened the skills needed by technologists, employers' recruitment and training policies often prevented technicians from entering the lower fringes of the technologist labour market.

At the other end of the spectrum the boundary between the technicians' and the craftsmen's labour markets is far from being sharply defined. A number of craftsmen have entered parts of the technician market through the extension of the content of their jobs. Manual operators have come to conduct routine laboratory tests on the plant; at one firm the job-title 'fitter' was changed to fitter-technician. New processes were making flexibility necessary and the fitters' job had been extended to embrace testing. Instrument mechanics had been promoted into the technicians' labour market through a general shortage of technicians and the need for the better utilization of labour. Technological change has widened the scope of manual workers' jobs and brought them into the lower reaches of the technicians' labour market.

At the two extremes it is clearly difficult to distinguish the technician's market from that of the technologist on the one hand

---

[1] This gives rise to the question of elasticity of substitution of one type of labour for another type. Employers' hiring standards can influence this elasticity, and hiring standards are affected by the availability of the type of labour required. The employer may want graduate employees, but being unable to obtain them he has to lower his hiring standards and take on workers whose qualifications are just below that of graduate status.

and the manual worker on the other; a further problem is the technician's perception of the external labour market, especially from the point of view of its potentiality for the satisfaction of his aspirations. An individual's perception of the labour market is influenced by his degree of knowledge of its structure, among other things. The more perfect the knowledge, the easier it will be to weigh up the costs and gains of movement. The degree of knowledge of the market acquired by technicians was difficult to assess. However, the research did show a regional variation in technicians' attitudes to the extent of the market. In the South-East the travelling of long distances from home to the place of work was seen as normal behaviour. In the North-West, however, commuting long distances to the work-place was thought to be abnormal. One plant in the North-West was within thirty miles of two large conurbations in Lancashire, but the technicians thought that the higher salaries which could be obtained in these towns would not compensate for the cost of the extra travelling. They had made little or no effort to find out whether this was in fact the case. Thus the size of the technician's labour market varies with each individual technician.[1]

Statistical problems in dealing with technicians' mobility

To assess the influence of the market on technicians' mobility some measurement of the state of the market is necessary. Unfortunately there is a lack of quantitative data on the supply-and-demand balance of technicians in relation to occupational requirements. The main statistical indicators of supply and demand in the labour market are the 'Wholly Unemployed and Unfilled Vacancy' statistics, published monthly by the Department of Employ-

[1] It could be useful to integrate A. W. Gouldner's concepts of 'Cosmopolitans' and 'Locals' in this respect. See 'Cosmopolitans and Locals: Towards an analysis of latent social roles', *Administrative Science Quarterly*, Vol. II, Nos. 1 and 2 (1957–58). *Cosmopolitans* have a low commitment to their firms but a high one to their skills, and tend to look to outside reference groups for progress comparisons. These groups would probably possess the greatest units of human capital and have the widest concept of the labour market. *Locals* have a high commitment to the organizations and make little reference to outside groups, and therefore are confined to the local labour market and unlikely to move into the international and national labour markets, although they may have the ability to do so. These concepts, however, are crude and require further refinements if they are to be integrated into a theory of labour mobility. However, the size of the individual technicians' market will be influenced by social and cultural factors.

ment.[1] The 'unfilled vacancy' figures may be taken as an indication of the demand for labour, while the 'wholly unemployed' figures are a guide to the potential supply of labour.

The occupational analysis of wholly unemployed and unfilled vacancies provide figures for laboratory assistants and draughtsmen on a national and regional level, but there are a number of limitations to the use of these statistics. The unfilled vacancy figures do not in reality represent the total number of vacancies, and fall short of the total numbers. Employers generally rely on other methods of finding technical personnel than through the employment exchanges, despite the establishment of the Professional and Executive Register. Employers in certain circumstances have a 'Standing Order' with the Employment Exchange to submit all suitable applicants to them without their 'notifying' any specific number of vacancies, and the vacancies remaining unfilled in such circumstances would not be included in the figures. Many of the firms participating in the study used the national press to recruit draughtsmen, industrial engineering personnel and quality-control services technicians. Some draughtsmen were on contract and hired through an agency; vacancies for draughtsmen are also advertised in the DATA Journal. Laboratory staff tended to be recruited directly from school or through the Youth Employment Service, but some of the more senior laboratory technicians were recruited through advertising in the local and national press. Despite these limitations, comparisons of the figures for various dates provided some indication of the change in the demand for labour.

The drawbacks of the unemployment figures have been well documented. The number of unemployed of all classes is published monthly and on an occupational basis in the quarterly enquiry into wholly unemployed and unfilled vacancies for adults by occupation. The important question is: how good are these figures as an indicator of the true amount of labour available to meet employers' demands? Inclusion in the unemployment figure depends upon an individual registering at an employment exchange. Many technicians will not register as unemployed, preferring to shop around for new employment or to use other

---

[1] An occupational analysis of wholly unemployed adults and unfilled vacancies for adults was for some time published quarterly in the *Gazette* and *Statistics on Incomes, Prices, Employment and Production*, but this source of information ceased in June 1969.

employment agencies. Many do not trouble to use the employment exchange because they consider that most technicians' vacancies are advertised in the national and local press. The figure for 'wholly unemployed' is drawn up on a particular day of each month, and to be included a technician would have to register at an exchange. The fact that many do not leads to understatement of the true figure. For technicians, then, the published unemployment figures are, at best, only a rough guide to the numbers that are available for any vacancies.

Having discussed the limitations of the wholly unemployed and unfilled vacancy statistics with reference to technicians, what can be said of them as indicators of the state of the labour market? Attempts have been made to use these series for the economy as a whole, and for certain industries to show trends in the demand for labour,[1] by constructing an ordinal index of excess demand for labour. Quarterly averages of unemployment and unfilled vacancies, seasonally adjusted, were expressed as a percentage of the total number of employees. This index is not a perfect measure of the strength of demand for labour, but makes full use of available data and is a better indicator than either unemployment or unfilled vacancies taken singly. However, the index was not used to try to assess the state of demand for a particular occupation, and, because of statistical shortcomings, it was impossible to attempt this in the present study. The Department of Employment and Productivity publishes an annual survey of occupations of workers in manufacturing,[2] but the technicians' group of occupations is given under the broad title, 'Other Technicians'. Only draughtsmen are shown separately. The unemployment and unfilled vacancy figures for draughtsmen are not given for manufacturing industry, but for all industries and services. However, the lack of occupational statistics for all draughtsmen employed in the economy makes the utilization of the Dow/Dicks-Mireaux index impossible. Unfortunately, at a lower level of aggregation, occupational statistics on a regional basis do not exist, although the unfilled vacancy and wholly unemployed by occupation figures are given

[1] J. C. R. Dow and L. A. Dicks-Mireaux, 'The Excess Demand for Labour', *Oxford Economic Papers*, February 1958.
[2] This series only began in 1963, and the 1969 survey was confined to the metal trades. This is to be the case with future surveys as part of the Government's acceptance of a request from industry that the amount of form-filling should be reduced.

on a regional basis. In this study, then, the cruder measures of unemployment and unfilled vacancies have been used to gauge the state of the market. Although these series have serious limitations, they are better than nothing. The balance of supply and demand in the market was judged by the crude difference between unfilled vacancies (v) and unemployment figures (u), so that where $u = v$ the market is in equilibrium, where $v > u$ the market is tight, and where $v < u$ the market is loose.

Appropriate national and regional data were obtained for only two groups (draughtsmen and laboratory staff). No statistical information concerning the local, i.e. town or city, labour market was available for any of the groups of technicians, and analysis at this level was on the basis of information obtained from the fieldwork.

The work situation

As a guide to the technician's satisfaction with the work situation, respondents were asked whether they had considered leaving their present job. Table 7.1 shows the replies from the questionnaire to this question. The overall picture showed that 52 per cent of technicians were not really thinking of leaving their present jobs. Less than 20 per cent definitely wanted to leave their jobs, but up to one-third showed some desire to make a change.

A detailed analysis of the characteristics of two groups of technicians, the majority group which expressed no desire to leave, and the members of the significant minority which would have liked to leave their jobs or were thinking of leaving them, was undertaken in order to isolate some possible factors explaining these attitudes. Because the majority of technicians appeared to want to stay in their present jobs this did not necessarily mean they were completely satisfied with their work situation. Some might not consider leaving because of the lack of alternative job opportunities within the limits of the extent of their perceived labour market. Secondly, mobility might not be a considered course of action for the technician possessing specific skills rather than general ones.

As a corollary to the question about the possibility of leaving, the technicians were asked to state the main reason for their answers. The main one appeared to be to obtain a post that would provide better career prospects. In all the occupational

Table 7.1 Technician replies to the question 'Are you thinking of leaving your job?' by occupational group

Occupational group	Yes, I am thinking of leaving this job	Yes, I would quite like to leave this job	Not really sure at all	No, I do not really want to leave my job	No, I am not thinking of leaving my job	Not applicable	Total replies from sample
Draughtsmen	20	11	13	21	33	3	100
Laboratory staff	12	17	16	18	35	2	100
Planning and production-engineers	17	19	20	13	24	3	100
Quality-control technicians	19	16	10	20	32	4	100

SOURCE: Questionnaire Survey.

NOTE: Figures are in percentages.

groups, better salary as a reason for leaving came out very low.
Of the reasons given for wanting to stay, satisfaction with the
present situation proved to be the highest for all groups. Security
of employment and acquired pension rights did not figure highly.
These results suggest that career development in the firm was the
most important factor in determining whether individuals were
thinking of leaving their job or staying in it.[1]

(1) *Attitudes to promotion outlets and desire to stay*

Studies of labour mobility have tended to ignore vertical mobility between and within departments inside a plant or a company,
and have concentrated on the movement from one employer to
another. Internal mobility is important. If the aspirations of a
group of workers to a higher position in the company hierarchy
of skills is blocked, frustration may arise, with consequences for

Table 7.2 **Reasons given for wanting to stay or
leave present work situation**

Reason for leaving or staying	Occupational Group				
	Draughts-men	Labora-tory staff	Planning and production-engineers	Quality-control tech-nicians	Total
	%	%	%	%	%
Leaving					
Better prospects	13	11	5	16	12
Better salary	4	2	5	5	4
Bored/under utilized	4	3	9	5	4
Personal reasons for leaving	5	8	5	7	7
Staying					
Pension scheme	2	3	1	0	2
Age	8	8	5	4	7
Satisfied	20	20	22	22	22
Personal reasons for staying	9	7	7	9	8
Security	3	2	0	3	2
No answer	33	32	40	28	33
Total	100	100	100	100	100

NOTE: Figures are in percentages.   SOURCE: Questionnaire Survey.

[1] The non-response for some broad occupational groups was high, e.g.
planning and production engineers 40 per cent, laboratory staff 32 per cent, and
draughtsmen 33 per cent.

the firm in morale and labour turnover. It was believed that promotion was a factor that highly motivated technicians, and answers to the questionnaire confirmed the hypothesis: 68 per cent of technicians were hoping to receive promotion to another job in the same field. Only 14 per cent of technicians said that they were not ambitious for further promotion, while 80 per cent indicated that they were anxious to achieve more promotion. The evidence from the questionnaire that technicians were generally aspiring and anxious to achieve promotion was further supported in plant interviews, where lack of promotion, and the phrase 'waiting for dead men's shoes', was a frequently heard complaint.

Management is clearly faced with the problem of providing outlets for promotion expectations. However, manpower planning as a method of avoiding promotion blockages is in an infant stage among British firms, and the majority of companies participating in the study had not made a forecast of their future requirements for technicians. Nor was there much awareness of the significance of manpower studies as an aid to the provision of a satisfactory career development for their existing employees.

All the firms in the study had formalized promotion structures within departments, but technicians thought the number of openings for promotion small. In some companies it was possible to move horizontally out of one department into another, and then be promoted upwards. Although this gave increased promotion opportunities the attitude of technicians in such companies to the number of openings available was still unfavourable. Aircraft firms appeared to be faced with a specific problem concerning career development. The industry was highly dependent upon government contracts, and while a contract was being worked it was easier for a company to accommodate its technicians' aspirations for promotion. However, when the contract finished, unless it was immediately replaced, further advance was limited. Promotion was possible only during the term of the contract.

There was in most firms a lack of promotion procedure whereby technicians could pass into technologists' jobs. In some of the companies it was deliberate policy not to allow anybody without a degree to move into technologist grades, and the 'graduate barrier' was seen by technicians and management, but for

Table 7.3   Technician attitudes towards promotion by broad occupational group

Occupational group	The statement is			
	True	False	No answer	Total
	You hope to receive promotion to another job in the same field			
Draughtsmen	71	22	7	100
Laboratory staff	67	26	8	100
Planning and production engineers	69	28	4	100
Quality-control services	66	28	6	100
Total	68	26	6	100
	You are not really ambitious for any further promotion			
Draughtsmen	18	80	9	100
Laboratory staff	14	8	5	100
Planning and production engineers	14	84	3	100
Quality-control services	20	74	6	100
Total	14	80	6	100

NOTE: Figures are in percentages.   SOURCE: Questionnaire Survey.

different reasons, as being difficult to cross. Some technicians had overcome the 'barrier', but these were few and were usually those who had been sponsored, by their company, on courses leading to higher educational qualifications.

Generally, vertical mobility within a firm was possible, but the extent of it appeared to be small, involving only a handful of technicians in relation to the total number employed. Some technicians had moved upwards, in that they had entered the company's employ as manual apprentices and then progressed on to the technical staff. But technicians considered there was insufficient opportunity for upward movement. Of draughtsmen replying they wanted to leave their job, 83 per cent considered promotion opportunities were insufficient, whereas of those who were not thinking of leaving only 47 per cent thought the opportunities for advancement insufficient. Of this latter group, 48 per cent strongly agreed or agreed that promotion opportunities were sufficient, but this was the view of only 11 per cent of the former group. Statistical tests found these differences in attitude to be significant.[1]

[1] Chi Square = 29.85; P < 0.0005. In order to calculate tests of significance in Tables the following categories were combined:
(1) 'Yes, I am thinking of leaving' and 'Yes, I would like to leave'.
(2) 'No I don't really want to leave' and 'No, I am not thinking of leaving'.
(3) 'Strongly agree' and 'Agree'.
(4) 'Disagree' and 'Strongly Disagree'.

Of the laboratory technicians answering that they were thinking of leaving, or would like to leave, 73 per cent considered
promotion opportunities in the plants in which they worked to
be insufficient. Only 25 per cent of these potentially mobile technicians were satisfied with their companies' promotion structure: the majority of laboratory technicians replied that
they would not like to leave, or were not thinking of leaving, and
of these 45 per cent considered promotion opportunities to be
insufficient. Thirty-four per cent of this group viewed the

Table 7.4 Attitude to promotion opportunities and readiness to leave job

Do you agree there is sufficient promotion?	Yes, I am thinking of leaving	Yes, I would like to leave	Not sure	No, I don't really want to leave	No, I am not thinking of leaving	Total
Draughtsmen						
Strongly agree	0	—	—	50	50	100
Agree	5	5	5	27	58	100
Not sure	10	3	10	24	53	100
Disagree	14	22	23	16	25	100
Strongly disagree	71	11	3	9	7	100
Total	21	13	15	18	33	100
Laboratory staff						
Strongly agree	9	15	19	16	41	100
Agree	10	3	26	26	35	100
Not sure	—	—	23	13	63	100
Disagree	12	18	18	16	36	100
Strongly disagree	24	24	24	22	6	100
Total	12	15	19	19	35	100
Planning and production engineers						
Strongly agree	—	—	50	50	—	100
Agree	15	5	10	15	55	100
Not sure	—	17	33	8	42	100
Disagree	25	17	9	16	33	100
Strongly disagree	25	38	13	19	6	100
Total	19	17	19	14	31	100
Quality-control technicians						
Strongly agree	—	—	—	—	100	100
Agree	—	25	8	21	46	100
Not sure	14	5	10	19	53	100
Disagree	20	16	16	22	27	100
Strongly disagree	41	20	10	23	7	100
Total	20	16	12	25	27	100

Figures in percentages.   SOURCE: Questionnaire Survey.

company promotion structure as adequate. Tests of significance[1] revealed that there were differences in the attitudes of 'potential leavers' and 'contented' technicians as to the sufficiency of promotion opportunities in the companies in which they were employed.

Significant differences were also found in the attitudes of quality-control technicians and planning and production engineering technicians as regards the sufficiency of promotion opportunities and the desire to stay or leave the present work-situation.[2] Of planning and production engineers who wanted to leave, 18 per cent regarded promotion opportunities as insufficient, and 20 per cent as sufficient. Of those who were not thinking of leaving the present work-situation, 45 per cent thought opportunities for vertical movement within their companies to be insufficient and 37 per cent sufficient. Of quality-control technicians thinking of leaving, 75 per cent were dissatisfied with the promotion opportunities, while for those expressing a desire to stay the figure was 50 per cent. Of 'contented' quality controllers, 25 per cent were satisfied with the opportunities for promotion in the companies in which they worked. The corresponding figure for potential leavers was $12\frac{1}{2}$ per cent.

Among the four main technician occupations, if a technician was considering leaving his present employer he was more likely to be dissatisfied with promotion opportunities than one who was 'contented'.

(2) *Attitudes to technician–shop-floor differential and potential mobility*

A further factor affecting the desire to leave or remain in the present work-situation was perhaps the attitude of these two groups of technicians to the relative wages in their company, and especially, the salary differential over manual workers in the same plant. It would be hypothesized that technicians with the greatest dissatisfaction over the shop-floor differential would be more likely to be thinking of or wanting to leave their present work-place than those who did not feel alienated by the width of this differential. It has been shown in chapter 6 that tech-

---

[1] Chi Square = 20.36; P < 0.0005
[2] For Industrial Engineering personnel, Chi Square = 10.93, P < 0.05, while for quality-control technicians the figures were chi. sq. = 9.79, P < 0.05.

nicians considered the technician–shop-floor differential to be too narrow.

The replies from technicians by occupation to the question 'Do you think that the pay of technical staff is insufficient, compared with the shop floor?' were shown in a previous chapter, Table 5.6. Eighty per cent of the respondents considered the differential to be insufficient and only 10 per cent thought it adequate. Among draughtsmen, approximately 85 per cent were unhappy about the differential, while the figure for laboratory staffs was 82 per cent. Eighty per cent of planning and production engineers, and quality-control technicians agreed that the pay of technical staff was insufficient compared with that of shop-floor workers.

There was strong feeling among all groups of technicians concerning the inadequacy of the technical staff–shop-floor differential. When an attempt was made to test the hypothesis suggested above, it was found that there was no significant difference in attitudes to the differentials between those technicians who were thinking of or would quite like to leave, and those who replied that they did not really want to or were not thinking of leaving their present job. Dissatisfaction with the width of technician–manual-worker differential did not appear to influence whether a technician was more likely to look for another job or to remain in his present work environment.

(3) *Skill utilization and potential mobility*

If a technician considers he is performing a task below the level of his training, frustration may build up and give rise to a situation of potential mobility. On the whole, among the four main technician groups skill utilization brought much or quite a lot of satisfaction; this was indicated, for example, by 63 per cent of draughtsmen, 49 per cent of laboratory staffs, 43 per cent of quality-control technicians and 58 per cent of planning and production engineers. However, more than one-fifth of the respondents in each group expressed dissatisfaction with the extent to which their skill was utilized in their work, e.g. draughtsmen 21 per cent, laboratory staffs 25 per cent, planning and production engineers 26 per cent and quality-control technicians 21 per cent.

A number of draughtsmen complained that they never saw

their work in operation and that draughting involved too much clerical work not requiring the qualifications of draughtsmen. In an electronics firm a laboratory assistant remarked: 'In the laboratory one is only called upon when needed and given no responsibility' and that, 'the skills acquired on day release courses are not used at work'. At an oil refinery a laboratory technician complained: 'So far as my job is concerned it is a

Table 7.5 **Sufficiency of pay of technical staff compared with shop floor workers, in relation to desire to leave job**

Do you think pay of technical staff is insufficient compared with that of shop-floor workers?	Yes, I am thinking of leaving my job	Yes, I would quite like to leave my job	Not sure	No, I do not really want to leave my job	No, I am not thinking of leaving my job	Total
Draughtsmen						
Definitely yes	22	14	9	23	33	100
Yes	22	12	20	19	27	100
Not sure	14	0	0	29	57	100
No	10	0	20	30	40	100
Definitely no	0	100	0	0	0	100
Total	20	13	13	22	32	100
Laboratory staff						
Definitely yes	18	14	19	22	26	100
Yes	8	20	8	16	47	100
Not sure	0	27	27	9	36	100
No	20	7	13	20	40	100
Definitely no	25	50	25	0	0	100
Total	14	17	16	19	34	100
Planning and production engineers						
Definitely yes	17	21	7	24	31	100
Yes	19	19	11	25	25	100
Not sure	66	0	33	0	0	100
No	9	5	41	5	41	100
Definitely no	0	0	0	0	0	100
Total	18	16	23	13	30	100
Quality-control technicians						
Definitely yes	20	13	17	27	27	100
Yes	26	13	15	26	19	100
Not sure	0	20	0	20	60	100
No	9	18	27	18	27	100
Definitely no	0	0	0	0	100	100
Total	20	13	15	26	26	100

NOTE: Figures in percentages.   SOURCE: Questionnaire Survey.

H

waste of time and effort to obtain an H.N.C. if I am going to remain with the company. A trained monkey can manage the job of laboratory assistant without straining its mental capacity.' This significant minority of technicians complaining of poor skill utilization gave rise to the hypothesis that technical staffs expressing dissatisfaction would be more likely to be thinking of leaving their present work situation than those contented with the utilization of their acquired skills.

The relationship between the desire of technicians to leave their job and the degree of satisfaction received from the proper utilization of their technical skills was not always as close as expected. For draughtsmen and quality-control technicians, tests of significance showed no difference in the attitudes to satisfaction or non-satisfaction from skill utilization between potentially mobile technicians and those who were thinking of remaining in their present work-situation. However, similar tests for laboratory staffs and planning and production engineers revealed significant differences in attitudes to satisfactions from skill utilization and the desire or non-desire to move from the present work-environment.[1]

(4) *Management policy and potential mobility*

Technicians' perception of management's policy is a variable that may help to explain the attitudes of those who do want to leave their present job and those who do not. In this respect, among the more crucial aspects of management policy are: the degree of interest technicians consider management to take in their problems; status accorded to technicians in the company hierarchy; and the grasp of technical problems by management.

By 'management' in this context was meant management at the plant, from departmental head upwards. Levels of management below this were excluded. A number of technicians argued that management was a lot more concerned with manual-worker trade unions than technical staff. Draughtsmen were equally divided over the degree of interest management showed in their problems. Eighteen per cent of quality-control technicians replying that they were thinking of or would like to leave their job, agreed that management took a reasonable interest

[1] Chi square (Lab.) = 31.04, P < 0.0005;
Chi square (Planners) 12.04, P < 0.025

in the technical staff of the company, while the figure for those who were not thinking of changing their job was 57 per cent. Sixty-six per cent of the potential leavers among the planning and production engineers disagreed or strongly disagreed that management took a reasonable interest in their technical staff, but only 25 per cent of those who implied satisfaction with their work situation criticized management interest in technical staff. The replies of draughtsmen showed the same trend. However, tests of significance revealed no difference in the attitudes of potential leavers and 'contented' technicians to the degree of interest they perceived management to take in their problems.

In the course of the fieldwork many comments were made concerning management's grasp of technical problems. At an electronics firm, one interviewee considered management in his firm to be too dominated by individuals with little technical knowledge, for example, accountants, while a laboratory assistant in an oil refinery thought his management was becoming too academic and that if the 'graduate barrier' were not removed, in a few years' time management would consist solely of non-practical technical people. In the aggregate, technicians considered management to have a weak grasp of the technical problems with which it was faced. Laboratory staffs who were thinking of leaving their jobs were stronger in their condemnation of management capability than those who said they were not thinking of leaving their jobs. Among planning and production engineers, 66 per cent of potential leavers considered management's grasp of technical problems to be bad or very bad, while only 33 per cent of those who were not contemplating leaving their present work-situation considered their managements to have a poor grasp of technical problems. The same trend was to be observed among draughtsmen and quality-control technicians. These attitudes might be linked to those of promotion opportunities. Since these were viewed as being insufficient, technicians may have regarded this as an outward sign that management did not value technical knowledge. But technicians saw technical knowledge as a vital necessity, and the failure of management to recognize this through promotion opportunities perhaps led some of them to consider management was unaware of technical problems.

The status accorded to technicians by management is a factor that gives rise to dissatisfaction and encourages job turnover. Technicians have strong views on the status which they believe is accorded to them by management. Forty-nine per cent considered the status of technicians to be too low in their companies, as opposed to 17 per cent who thought their status duly accorded. Among draughtsmen only 12½ per cent considered their status to be high enough in the company, and the corresponding figures for other groups were: laboratory staff approximately 20 per cent, planning and production engineers 20 per cent, and quality-control technicians 14 per cent.

Technicians tended to see the relative position of manual workers, rather than technologists, as the critical reference group with regard to their own status. Many technicians complained of their salary declining in relation to that of manual workers, and considered this to be evidence of declining status. Other technicians saw another outward sign of declining status in one of the companies where a job-evaluation scheme had placed lower-grade technicians in the same grade as clerks. One laboratory assistant remarked, 'Generally, the technical staff are in a poor position in this establishment, being considered equivalent to clerical staff, whose responsibilities and qualifications are in fact lower than those of any laboratory technician.'

At another plant of the same company, a laboratory technician explained: 'I feel I belong to a section of industry becoming submerged. We are being drowned by highly paid and highly qualified graduate staff, and also by highly organized but poorly qualified manual workers who belong to powerful unions. There are too many foremen doing jobs we could do. Automation also threatens many of us.' Elsewhere laboratory staff argued that the status of technical staff in society as a whole was too low. Technical staff, it was argued, were definitely without status, unlike doctors and solicitors, and this should be changed. Senior engineers, for example, should be given a different title—say 'engineering managers'.

The apparent discontent with the status accorded by management, together with the fact that a small minority considered their status to be satisfactory, led to the hypothesis that a technician considering a move from his present work environment would be more likely to be one who was dissatisfied with the status accorded to technicians by management than one who was satisfied with the status of technical staff. However, statistical tests showed that there was little difference between the two groups in terms of their perceptions of management's concern about the status of technicians.

The results of the statistical tests of the factors that were thought most likely to induce technicians to change their jobs are set out in the table below.

Table 7.6 **Differences in attitudes to certain plant variables between leavers and contented technicians**

S = significant   NS = not significant

Occupational group	Career-development	Relative wages	Skill utilization	Management policy	
				Interest in problems of technicians	Status
Draughtsmen	S	NS	NS	NS	NS
Laboratory staff	S	NS	S	NS	NS
Planning and production engineers	S	NS	NS	NS	S
Quality-control technicians	S	NS	NS	NS	S

SOURCE: Questionnaire Survey.

On the basis of this evidence, a number of tentative propositions can be made.

1. For all the main four groups of technicians' occupations, if a technician is thinking of leaving his present work-place he is likely to be more critical of promotion opportunities than one who is not.

2. For draughtsmen, the main variable in explaining the decision to leave is the lack of career development in the present company.

3. Laboratory staffs who are thinking of leaving their employers are likely to feel more strongly about the lack of career development and skill utilization than the laboratory staffs who desire to stay within their present work-environment.

4. If planning and production engineers and quality-control technicians are considering moving to a new job, they are more likely to be more critical of the lack of career development and the status accorded by management to their occupations in their companies than those who are not contemplating a change of employer.

5. The attitudes of all the four main occupational groups to the insufficiency of the technician manual–worker differential and the lack of interest by management in the problems of technicians do not appear to be significantly different between those contemplating mobility and those wishing to stay with their present employer.

Factors working against mobility

1. *At the work-place*

There are factors at the work-place discouraging mobility, and among these can be included age, length of service and the type of skills possessed by the technician, i.e. general or specific. Up to a certain age, it is probable that a technician will have made no firm decision about staying at this present company. After some critical age has been passed, individuals are more likely to have made a definite decision about the duration of the present job they are holding. If a technician is under thirty years of age, under contemporary conditions he will be more likely to be thinking of leaving his job than a technician who is over thirty. Of draughtsmen under thirty, more replied that they were thinking of changing their job than that they were not thinking of leaving. The same was true for laboratory staffs. As with draughtsmen, among planning and production engineers, those under thirty years of age expressed a stronger desire to change their job than those over this age, but statistical tests showed that there was no significance between the two groups as regards potential mobility and non-movement. With quality-control service technicians, thirty again was a dividing line between the desire to leave being stronger than the desire to stay. However, this difference between the two age-groups was only weakly significant.

Length of service can influence a technician's attitude to whether he wants to change his job. Technicians who have only a short period of service, say under two years, will probably not want to change their job because the trial period in that job has not yet expired. The replies from the sample supported this view, except for quality-control service technicians. Of those who said they were thinking of leaving their job, nearly 50 per cent had been on the technical staff of their companies for less than two years. A processing of the sample response revealed that the length of service became a factor working against mobility after a technician had been on the technical staff of his company for six years or more.

If a technician was moving from one work-situation to another, the chances were high that he had been on the technical staff of his present company for between two and five years. It is reasonable to expect that the longer a technician has been on the staff payroll of his company, the less likely he is to want to leave his job, since he may lose his accrued pension rights. Additionally, the longer the technician stays in a firm and its attached external community, the tighter will become his links to the area, so that external sociological factors will help to prevent mobility.

The possession of specific skills can be an obstacle to mobility not only between firms, industries and regions, but between departments within a plant or company. Specific skills are those only usable in the firms that provide work requiring those skills and are of little use in other firms, occupations or industries. In the two chemical plants, technicians employed in technological departments were receiving specialized training, and the acquired skills were not easily transferable to other departments in the refinery. In one electronics firm, there were a number of testers and inspectors doing routine tests such that the skills acquired were of little use for other jobs, while in a vehicles plant, road-test driver-technicians were carrying out work of a limited specialized nature on the testing of brake linings. In the trials engineering section of an aircraft firm there were technicians concerned with supervising tests on aircraft instruments. The manager of the section was aware of the specialized nature of the work, and the problem presented for vertical mobility within the firm, since transfer to other departments was impossible. All the groups mentioned above were restricted in the amount of mobility they could undertake since demand for their skills was limited, and movement might only be possible if the labour market was perceived in its widest extent. For the lower-grade technicians, for example testers and inspectors, their skills were so specific that even if such technicians viewed the world as the extent of their market, movement would be difficult. Technicians in oil-refinery technological departments would easily find jobs in other oil refineries, but this meant geographical mobility, and such movement might not be undertaken because of certain external factors, for example, ties to the area where they resided.

General skills are those which can be used in any firm, and technicians with such skills have a greater ability to move than those that possess specific skill. For technicians with general skills the main constraints to geographical mobility come from outside the place of work. Mobility can frequently take place within a firm, between employers, and between industries, but it may be confined to a local geographical area. Hence job opportunities within the technician's perception of the labour market will be an important external determinant to mobility. Within the study the occupational groups with general skills were draughtsmen and planning and production engineers, as well as some laboratory staffs. Draughtsmen had skills in maintenance, mechanical and civil drawing, and the demand for each of these types of draughtsman was high. There were, however, exceptions, and a group of draughtsmen in one firm in the North-East had little ability to move because of their lack of formal qualifications and relatively high age. The general skills of the draughtsmen were perhaps best illustrated by the growth of self-employment among this group. Demand for their services was common to employers in most industries, and this, along with the excess demand for draughtsmen, enabled the practice to grow.

(2) External to the work-place

Economic and social variables external to the plant may work against mobility. If a technician is dissatisfied with his present work-situation, before a move to a new situation which may involve geographical mobility is undertaken, he may attempt to weigh up the costs of movement to a new work-place against the loss from leaving the present community within which he lives. If an individual considers the benefits he gains from living in his present community outweigh the disadvantages of his present work-situation, then movement may not take place. From interviews examples were found of ties to an area being too strong to break. In a metal-manufacturing company in the North-East many of the technicians said they would like to change jobs but would not like to leave their present community. Reasons given included family ties in the area and difficulties in selling houses for a sufficiently high price to buy property in a more expensive area. The phenomena of social commitment to an area acting as a brake to mobility was also found amongst planners and production controllers at an electronics firm in south-west London. Here technicians mixed more with other groups at the work-place and had ties with the company as a community since they had grown up with the firm. This situation contrasted sharply with that of testers and inspectors in the same company, where there was a high turn-over and little commitment to the firm and community.

There were draughtsmen and technicians who were not prepared to move, appearing to have little ambition to do so. It is possible that this attitude had been fostered as a result of the experience of the present generation of technicians' fathers in the depression of the 1930s. For such people employment had been difficult to find, and this may have influenced their children into thinking that to have a job was fortunate, and if one had one it was better to keep it and not risk the unknown.

The economic constraint on mobility applies not only to geographical mobility but also to movement between firms, occupations and industries within a local labour market. Movement will be determined by the technician's perception of the labour market, which in turn will be influenced by his degree of knowledge of the market, and the size of the labour market for the individual; for example, some may be prepared to move only within a local radius, but the relevant area for others would be the regional, national or international labour market.

Actual movement

So far this chapter has concentrated on potential mobility by looking at differences in attitudes to certain variables among technicians who had expressed a desire to leave and those who had no wish to leave their present work situation. But discontent with the work situation does not necessarily lead to mobility; and an attempt was made to find the extent to which technicians had actually changed jobs. The personnel departments of each company were asked to delete the names of those technicians who had left the firm's employ twelve months after the visit of the research team. Of the establishments participating in the study twelve sent back usable replies. The non-response company explained that since the research in their plant, special circumstances had led to a large amount of mobility which it considered would have given a false impression of normal turn-over and recruitment requirements.

(1) *Past mobility*

Draughtsmen had experienced most mobility in so far as 60 per cent had been recruited to their present job from another firm. Planning engineers, who were generally older men, were not so likely to have changed their jobs already: 45 per cent had joined their present employer from another company. These technicians (i.e. planning and production engineers) could be divided into the highly qualified senior men, often ex-draughtsmen, who had made a move in the course of gaining upward promotion from the drawing office, and the 34 per cent of ex-craftsmen who had been promoted within the firm without moving.

Table 7.7 Source of recruitment of the technician labour force

Occupational group	Joined from school	Promoted while doing a craft apprenticeship	Promoted on finishing craft apprenticeship	From a non-staff job	Another white collar job	Another company	Manual	Other	Total
Draughtsmen	9	8	10	7	3	60	11	13	100
Laboratory staff	40	1	1	12	4	25	2	13	100
Planning and production engineers	8	5	5	23	8	45	1	3	100
Quality-control services	6	2	2	24	3	48	1	16	100
Overall Total	19	4	5	15	4	43	1	9	100

NOTE: Figures are in percentages.   SOURCE: Questionnaire Survey

Quality-control technicians were nearest to draughtsmen in terms of prior mobility, 48 per cent having been recruited from another firm, but they were, like planning and production engineers, unlikely to achieve further upward movement between jobs, if only because they too were older men. Again there were two clear groups: the more qualified controllers doing test jobs in electronics or the chemical industry, and the ex-shop-floor inspectors with experience but few formal qualifications. Although these groups bear the same occupational title, they are in different labour markets. Laboratory assistants were lowest in terms of previous mobility. Only 25 per cent had joined from another firm, and for a much larger group (45 per cent) their present job was their first and only experience in the labour market. Those who had moved could be categorized into those with high formal qualifications who might well have to move in order to obtain promotion, and those with O levels who could not easily move again to their own advantage without further qualifications.

(2) *Present mobility*

(a) *Overall*. The returns from the companies showed that 14 per cent of technicians had left the employ of the company in the twelve months following the research. The figure included technical staff who had moved between departments within an establishment. The occupational breakdown is shown in Table 7.8.

Table 7.8   Mobility among technicians between June 1968 and June 1969

Occupational group	Number moved	Number in sample population	Percentage turnover rate
Draughtsmen	53	300	18.0
Laboratory staff	32	315	10.0
Planning and production engineers	48	262	18.0
Quality-control technicians	37	221	17.0

SOURCE: Personnel records of companies participating in study.

The highest turnover was amongst planning and production engineers, draughtsmen and service engineers, and lowest amongst laboratory staff. For the other groups, the figure ranged between 12.5 per cent and 18 per cent. On the returned sample lists some employers indicated where some of the technicians had moved, but this was by no means general. Some draughtsmen had moved from their department for internal promotion, others had gone into contracting work, and some into local government and teaching. The tight labour market for technicians enabled the ambitious draughtsmen to leave their present firm and gain experience elsewhere. Unfortunately it was impossible to identify the sub-groups to which the technicians who had moved belonged. It is more than likely that the draughtsmen and laboratory staffs who had moved were the most highly qualified amongst their group. Among the planning and production engineers, it was the highly qualified, more senior men, who had been ex-draughtsmen and who saw movement from employer to employer necessary to gain all-round experience, who had changed jobs.

(b) *Draughtsmen.* In vehicles and engineering the turnover of draughtsmen was over 20 per cent. The low turnover in the chemical industry is to some extent explained by the high rates of pay in this industry in relation to others (see Chapter 6). The metal-manufacturing figure had been affected by a rumour of closure of one of the firms, while the vehicles figure was inflated, partly because of fear of the consequences of a proposed merger between the company concerned and another firm. The plant was in an area where the results of the G.E.C./A.E.I. merger provided an example of how technical staffs could be affected by rationalization moves.

An analysis of the movement of draughtsmen on a regional level shows that the figures for the South-West region were well above the average, as a result of a redundancy in one of the firms in the region soon after the research had been completed. It was an area where the market for draughtsmen was narrow After allowing for this, the turnover of draughtsmen was highest in London and the South-East, an area where alternative job opportunities were greatest, and this was liable to give rise to voluntary mobility. The figure for the Northern region was high in relation to market pressures, which were slacker than in the other three regions for the period 1964–68, but was inflated by the fear of corresponding redundancy by a large employer undertaking 'rationalization'.

(c) *Laboratory staff.* The turnover rate of laboratory staff was lower than that of draughtsmen. The industry showing a contrary trend was engineering and electrical goods manufacture, which had a turnover rate of over 70 per cent. Some of the figure was accounted for by internal promotion and a large movement of laboratory staff in the North-West region into school-teaching.

The fear of redundancy in one of the firms in this industry in one part of the country also helped to drive up this figure.

On a regional basis turnover amongst laboratory staff was highest in the South-West and North-West and lowest in the South-East and Northern regions. The South-West figure was biased in that one of the companies in this region declared a redundancy amongst its technical staff shortly after the completion of the research at the plant. The North-West figure was accounted for by a large turnover among laboratory assistants in the engineering firms into school-teaching. The low turnover of laboratory staff in the Northen region could be related to the state of the labour market, since in this area the availability of laboratory jobs was liminted in relation to that in the other three areas.

(d) *Planning and production engineers.* The planning and production engineers who had moved were most numerous in vehicles and least in chemicals and metal manufacture. In the South-West the occupational group showed an unexpected movement, which was the result of the previously described redundancy in the region. Turnover of industrial engineers was highest in London, the area where employers' complaints of a shortage of labour were greatest.

(e) *Quality-control technicians.* The turnover of quality-control technicians was greatest in vehicles and least in metal manufacture. However, to see whether mobility was linked to the state of supply and demand for quality-control technicians, it seemed better to examine movement on a regional basis. The turnover amongst quality-control technicians was highest in London and the South-East, and the North-West. In both these areas, job opportunities for inspectors and testers were plentiful. The higher figure for the South-West was explained by a redundancy that took place in one company after the research.

Conclusions

The 32 per cent of the sample who said they were thinking of changing their present employers exhibited a dissatisfaction with promotion opportunities within their firms which was significantly greater than that of those who had no intention of leaving.

While both groups showed great concern over the narrowness of the technical staff–shop differential, no significant difference was found between them. There was no significant difference between 'contented' and potentially mobile groups of technicians in their concern about the degree of management interest in the problems of technicians. Only among laboratory staffs was there any significant difference between 'leavers' and non-leavers in their attitude to the degree of satisfaction derived from management's utilization of their skills. For planning and production engineers and quality-control technicians, there were significant differences between 'contented' and 'estranged' technicians as to the degree of status accorded to them by management.

An examination of factors within the firm that might work against voluntary mobility was undertaken. Age and length of service were found to be significant among certain occupational groups in that those over 30 years of age and who had had six or more years' service were less likely to consider changing their jobs than those under 30 years and who had had less than six years' service. Of factors external to the plant working against mobility, ties in the community were not found to be important except in the north-eastern part of the country.

A study of the actual mobility of technicians showed this to be small. Only 14 per cent of of technicians had changed their jobs in the twelve-month period following the research in the plants concerned. Mobility was highest amongst planning and production engineers, and lowest amongst laboratory staffs. The most significant (though perhaps obvious) conclusion which emerged was that, although in some cases special factors tended to interfere, movements were greatest where the labour market was tightest.

8 Influences upon Attitudes to Unions and Professional Membership

In Chapter 3 the institutional factors affecting the processes of unionization and professionalization among technicians were discussed. The examination of the historical growth-path of these organizations suggested that groups formed around a common concern with status as well as with market position. Draughtsmen had discovered an occupational identity at the end of the last century; they had quickly acquired prestige and commanded a scarcity rent in the labour market. In the last seventy years the status and market value of draughtsmen have been affected by the explosive growths in demand caused by the two world wars and the great expansion of modern industry and its attendant technology. But hopes entertained for the creation of professional occupation sustained by the device of strictly controlled entry faded with the changing structure of industrial organization and the pattern of demand for draughtsmen.

The labour-market and work situations of laboratory and production technicians were more heterogeneous than for draughtsmen, making it even more difficult for them to generate the social cohesion necessary to establish strong professional associations.

The problems which faced technicians' organizations were complex. The effort to win and to serve members effectively had to take into account differences in the response of technicians to a variety of technological and market situations. In their attitude to membership of trade-union organizations, technicians were influenced by the fortunes of comparative reference groups, especially manual workers immediately below

and technologists immediately above. The fact that they were
employed in small groups, often had a close contact with higher
management, and aspired to promotion, placed them in an
ambiguous position. Where their market position was strong
they had no need for collective bargaining since they had sufficient power to engage successfully in individual bargaining.
Yet in spite of the normal reluctance of technicians to join unions,
they did so spontaneously under sustained stress, or where it
was locally accepted and appropriate to do so.

The market strategies of both draughtsmen and the other
technicians' associations moved towards a greater acceptance
of collective bargaining on behalf of 'all working at the tools'[1]
and away from attempts to create closed professional associations. Although the influences of changing market and technological factors are not to be discounted in this process, the
ideology of the associations' leaders and that of their activists at
local level was almost certainly a decisive factor in bringing
about the institutional changes in strategy.

A strategy built on a use of differential status to secure monopolies in the market by the restriction of entry,[2] and upon
the use of a status ideology to attract members or 'to shape the
character of organization'[3] was a viable *complement* to that of a
'class' or 'conflict' strategy of organization and *not an alternative*.
This situation appears to contradict the notion that organizations may be classed along a continuum of institutionalized
characteristics such as is suggested by the polar types of the
professional association and the 'open' union.[4] At a time when
DATA was experiencing its most militant phase under a leadership attempting to encourage feelings of solidarity with the
shop-floor members of the Amalagamated Engineering Union, the

[1] See Sidney and Beatrice Webb, *Industrial Democracy* (Longman, London
1919).
[2] To quote: 'As to the general effect of the status order, only one consequence
can be stated but it is a very important one: the hindrance of the free development of the market occurs first for those goods which status groups directly
withhold from free exchange by monopolisation. From this monopolisation
may be effected either legally or conventionally.' *From Max Weber Essays in Sociology* O.U.P. New York, 1946. H. H. Gerth and C. W. Mills (eds.), *Essays in
Sociology* Chap. *VII*. (New York: Oxford University Press, 1946).
[3] David Lockwood, *The Black-coated Worker* (Allen and Unwin, London
1958).
[4] See R. M. Blackburn and K. Prandy, 'White Collar Unionisation: A
conceptual framework' *British Journal of Sociology*, Vol. XVI, No. 2 (June 1965).

General Secretary was justifying the unions' pay policy as follows: 'If [the draughtsman] had chosen to remain in the workshops in a manual craft capacity—not having to accept the higher demands for technical education, performance or responsibility, then at the age of twenty-one years he would enjoy earnings several pounds more per week than had he accepted "promotion" or "creaming off" into the design office.[1] This comment, and numerous similar leadership statements, were calculated to receive a sympathetic hearing among an audience believing themselves deprived of the just and appropriate rewards of their 'calling' or vocation in the drawing office. Mr Doughty's remarks also imply that the feeling of relative deprivation may be felt on a generalized occupational basis which extends beyond that of the immediate work environment.

In this chapter we explore the degree to which attitudes displayed in replies to our survey coincide with those suggested by the institutional characteristics analysed in Chapter 3. In other chapters we have already commented on the discontent arising from market and work situations in which technicians' expectations for promotion and desire to identify with management were constantly frustrated. There appears to be an increasing disjuncture between their expectations and their 'objective' position. All groups displayed a high degree of insecurity, but this did not necessarily lead on to any universal consciousness of a shared market position. This wider identity usually existed only where shared experience in terms of similar training or education had been reinforced by the isolation of large numbers of technicians on the basis of their function within the work process. This was especially true of draughtsmen, to whom Mr Doughty directed his comments, and also for many laboratory technicians. But for these, as for other technicians, more dispersed in their work situation, the issues most clearly felt were expressed in relation to local status hierarchies.

The expression of grievances through unions depends upon a decision to join unions. The need to make this decision and to decide whether this is a form of behaviour 'appropriate' to one's status position is in fact usually only required at some historic point of time when the union is introduced into the

---

[1] *Royal Commission on Trade Unions and Employers Associations*, Minutes of Evidence No. 36, DATA (H.M.S.O., London 1966).

particular work situation.[1] After that time it becomes one of the three accepted and available collective means by which technicians may express a *collective* grievance; the others being those of the staff associations or professional associations. The choice of a trade union or professional association did not however appear to indicate any deep sense of commitment to a wider class or occupational identity, nor did it necessarily even indicate any great awareness of a similarity in market position with groups outside the immediate work-environment. Personal affiliations and current organization linkages appeared to be viewed in an 'instrumental' manner which led technicians to accept whichever strategy appeared most effective in meeting their goals.

Our findings must be regarded as extremely tentative, since, because of operational difficulties, we were unable to explore a variety of political and community aspects of the technician's self-image in our questionnaire sample.[2] However, the manner in which technicians accepted both professional bodies, and displayed a task-oriented concern for intrinsic aspects of job satisfaction irrespective of union membership, while all the time displaying a great concern for their market position in relation to that of manual workers, suggests that a simple dichotomous view of such institutional affiliations is wholly misleading.[3]

Technicians and unionization

Of the four main occupational groups upon whom this analysis is concentrated, only laboratory assistants had a majority of

[1] This kind of group choice occurred, for example, in three of the more densely unionized plants in our fourteen-plant study, in which spontaneous strikes led to the introduction of ASSET into the work situation.

[2] Questions relating to political and other affiliations were removed at the request of one of the participating trade unions. This was not done until after our initial sample in a South Wales steel plant had revealed no significant effect from such affiliations. Technicians selected for interview answered questions relating to their own and their members' wider beliefs. Indeed, it was sometimes necessary for interviewers to cross-check the apparently exaggerated comparisons made by union representatives of their members' political *apathy* with their own (i.e. the activists') high level of *commitment*.

[3] The essentially static descriptive continuum used by R. M. Blackburn and K. Prandy in 'White Collar Unionization: A conceptual framework in *British Journal of Sociology*, Vol. XVI, No. 2, June 1965, implies a similar dichotomy. This should be contrasted with the dynamic model put forward by the Webbs (in *Industrial Democracy*), in which concern for local status is seen as an initial stage on the way to collective bargaining.

trade unionists. Draughtsmen, who are generally thought of as the most unionized technicians, were much less so in our sample.

Table 8.1 Are you a paid-up member of a trade union for technical staff or of a staff branch of a predominantly manual workers' union?

Membership of a union	Laboratory staff	Draughtsmen	Planning and Production engineers	Quality controllers
	%	%	%	%
Yes	56	43	22	37
No	40	49	58	57
No answer	4	7	22	6
Total	100	100	100	100

NOTE: Figures in percentages. SOURCE: Questionnaire Survey.

The most important difference between union members among laboratory staff and draughtsmen on the one hand and among planning and production engineers and quality controllers on the other was that of age. The former group was generally younger than the latter, but only among laboratory assistants were the differences in membership between age cohorts statistically significant. In the laboratories, technicians under the age of 25 were far more likely to be in unions than those over 25 ($x^2 = 1.044$ P$<$0.05). With these differences in age there were associated differences in school attainment and level of formal qualification (though these differences were not all in the same direction).

Of laboratory assistants under 21, 82 per cent were trade unionists, and a majority of draughtsmen unionists were under 30. These young people were less likely to be in possession of some formal qualifications in their work than the somewhat older unionists in other technical occupations, or indeed, than their older laboratory colleagues who were not in unions. The significance of this fact will be brought out later.

The second major group of unionists was that of quality controllers and planning and production engineers in the 31–40-year-old cohort. They had usually gained a good school qualification, followed by the attainment of at least a City and Guilds Certificate.

Technicians belonged to a range of organizations, but

draughtsmen were always members of the Draughtsmen's and Allied Technicians' Association, and the majority of laboratory assistants were in the Association of Scientific Workers. For nearly two-thirds of current union members among the young laboratory assistants their only experience of unions was their present one, whereas this was true of only 40 per cent of informants in the other occupational groups.

The numbers of technicians who were currently in trade unions was not the only guide to the propensity to join unions; a good many non-unionists had at some time been members of a union, whatever their present status. This was true of 73 per cent of laboratory assistants, 70 per cent of draughtsmen, 48 per cent of quality controllers and 45 per cent of planning and production engineers, all of whom had once belonged to a technicians' staff union. This implies a high union turnover rate in three groups, particularly among draughtsmen and laboratory assistants, most of whom at some stage in their career had been in a union, though only for a short period; a more stable membership record was found among quality controllers.

The reasons given for leaving unions were varied, but it was clear that for many technicians the national policy of the unions and the ideological stance of their leaders at local as well as national level had an abrasive effect on their attitudes. Typical comments from former members were as follows:

'I believe that [your] questions on trade unions in the technical field were unnecessary because there is absolutely no need for technical union representation. Technicians do and should advance on their own merits and not in large classified groups.' This comment came from a quality controller in an electrical goods plant employing large-batch production methods. He had previously spent half a page in complaint about the lack of promotion opportunities in his field of work. A comment which represented a body of opinion in the drawing office came from a draughtsman recently promoted to section leader in an electronics plant:

> there was a time when I thought that the unions were a fine thing, but of late years the irresponsibility of the office committees have forced me to resign. In our own office they can hardly be said to reflect the views of the members, even the members agree on that! Yet they are too apathetic to do anything about it. Further, the Unions have formed a barrier between the workers and manage-

ment ... one more point, membership of a Union, especially that applying in the office, *infers that such a member is qualified to be a draughtsman,* though having no training or technical background!

This display of resentment suggests that to many technicians the existence of a union is evidence of the breakdown of a relationship to work and to management which they valued highly. Management disapproval could make a union 'taboo' for some technicians, while approval could make it acceptable and 'appropriate' behaviour to belong. There is a strong suggestion within the survey data that the technician's view of union membership depends, *inter alia,* on his career position and the concomitant work situation in which he finds himself. That is not to say that the ideological stance adopted by the draughtsman section head was necessarily adopted *ex post facto,* but rather that it became manifest only when attachment to the 'management team' established his superior work-status. There were several active unionists in the sample in supervisory positions. These people required a high 'moral'[1] commitment to the union since most managements refused to recognize it as representing other than the most junior grades.

Within the work situation the feelings of the membership group to which the technician belonged was extremely important. To join a union in the face of opposition from his peers required a commitment normally found only in an ideologist, or one whose normative reference group lay outside his work environment.[2] For this reason he is singled out in this analysis as a 'pioneer'; in several plants the continued presence of the union within the work environment depended on such a person.

Alternatively, if the majority of the work-group to which the technician was a new recruit were in a union, he required a very strong belief in 'individualism' to hold out for long against such group pressures. This point will be examined in depth later in the chapter.

Professionalization

The level of membership of professional associations is low when compared with that of trade unions (Tables 8.1 and 8.2). This

[1] A. Etzioni, *A Comparative Analysis of Complex Organisations* (Free Press of Glencoe, New York 1961).
[2] For a fuller description see C. Wright Mills, *White Collar: The American Middle Classes* (New York: Oxford University Press, 1951).

is not surprising. Indeed, when one considers the reliance of the professional institution on an initial approach by the individual to the institution, as compared with the proselytizing nature of union recruitment, the figures appear extremely high.

The most important reason for not regarding membership of professional associations as a measure of propensity for formalized collective action equivalent to that of trade-union membership lies in the nature of the organizations themselves. To *acquire* membership of a professional body it is normally necessary to have first acquired a formal qualification by examination. While some of the trade unions organizing technicians retain a measure of 'closure' through the criteria used in recruiting members, none of them are anything like as selective or erect such barriers to entry as those used by professional associations. In order to make a proper assessment of the attraction of these bodies, it is perhaps better to measure the propensity of those with appropriate qualifications to join them.

In addition to these self-imposed limits to membership, 'professionalization' among employees may be actively discouraged by managers. The constraints upon management authority represented by membership of any external work organization have already been suggested. Differentiation of skills through professional qualifications acts upon the elasticity of the supply of labour available to the entrepreneur. Recognition of a specific expertise in a work area diminishes the ability of a manager to issue instructions in the area. To the degree that management are willing to pay more for a diminished labour supply and to recognize the autonomy of a 'professional technician' they will encourage membership.

In the case histories it was discovered that, as with membership of trade unions, membership of professional institutions was highest where it was encouraged by first-line supervisors. Management above first- or second-line tended to be antipathetic to membership of professional organizations. Training within the company was regarded as a purposive task-oriented operation. At the same time, management views on recruiting from *outside* the company offered a paradoxical contradiction. For external recruitment purposes, formal qualifications, including membership of reputable professional associations, were of rapidly growing importance.

Technicians placed a high value on work autonomy, and, as

Table 8.2   **Are you a member of a professional institution or association
for technical staff? Percentage replies**

Occupational group	Total number of replies	Number not in professional body	Number in professional body
Laboratory staff	100	83	17
Draughtsmen	100	87	13
Quality controllers	100	95	5
Planning and production engineers	100	70	30

SOURCE: Questionnaire Survey.

was shown in Chapter 4, were highly interested in the possibility
of new professional qualifications for technicians. Yet given this
ambivalence on the part of management it is not surprising that
the more aspiring technicians and those who see their outlook
as having more in common with management are not necessarily
those who actually join professional institutes.

Of the four occupational categories upon which our analysis
is concentrated, membership of professional institutions was
greatest among planning and production engineers: 30 per cent
of this group were members of some institution. This group
contained a wide span of formal qualifications, from 5 per cent
graduates to 11 per cent whose sole qualifications were those of
their professional institution. But relatively speaking they were
technically well qualified, particularly among the younger technicians in this category (26 per cent O.N.C., 28 per cent
H.N.C. among the under thirties). In many ways these people
stood out as being typical of the professionally aspiring technician in industry. Many of the planning and production
engineers were older men, who had good reason to regard their
present positions to which they had graduated through the
Drawing Office as terminal within their career, but rarely
brought themselves to say so. The work-study engineer was often
a man promoted from a manual craft or supervisory job who was
inclined to lay great emphasis on the 'scientific' nature of this
job and the need to set professional standards.

A second and increasingly important group of informants
was composed of young men recruited fairly early in life who,
while being interested in examinations as a means to an end,
did not regard professional institutions as necessary to assert

their status or to identify and define their role. College qualifications and their own ability were seen as being more effective. There was a lack of association between the possession of technical qualifications and the expressed need for professional institutions, which became strongly negative among under thirties in the laboratories and drawing offices.

By way of contrast, among the least formally qualified, instrument technicians (often members of the Electronics and Electrical Trade Union), and among company-trained draughtsmen, the need for professional bodies received greatest support. In other words the younger and more confident technician was much less enamoured with the existing professional institutions than was the older, in terms of the lack of formal certifications, the less qualified man.

With so many blockages (including lack of knowledge) to membership of professional associations, it seemed better to measure the propensity for membership of professional institutions rather than the achieved membership. The major relationships to be examined in this chapter are therefore those between the major factors which structure career opportunities such as education and qualification, and a propensity for professionalization. Answers to the question 'Does membership of professional bodies seem irrelevant for technical staff like you?' (see below) are treated as displaying a propensity (or not) for professional membership; technicians perceiving professional institutions as a relevant factor are referred to as 'professionalized'.[1]

'Unionized' professionals and 'professionalized' unionists

It is quite clear that many technicians do not see any contradiction in regarding themselves as both 'professionals' and 'trade unionists'. Of the 35 laboratory staff in professional institutions, 15 were in unions and 11 did not give an answer to the question on unions. This proportion (43 per cent) of quality controllers and planning engineers were also in unions, but the proportion fell to 20 per cent among draughtsmen. The majority (52

[1] The immediate operational practicality of this usage of the terms 'professionalized' and 'professionalization' will be immediately apparent in view of the small proportion of our sample population who were in professional associations. But it is argued here that in view of the 'closed' nature of the institutions the propensity or the *process* of professionalization is more important than *form*. If a technician feels a *need* for professional membership this is regarded as the significant factor to be examined.

Table 8.3 Does membership of professional bodies seem irrelevant to technical staff like you? – Percentage replies

Occupational group	No answer	Most irrele-vant	Quite irrele-vant	Not sure of its relevance	Quite rele-vant	Most rele-vant	Don't know	Not appli-cable
Laboratory staff	8	14	16	27	15	11	6	4
Draughtsmen	8	9	22	17	25	15	—	3
Planning and production engineers	22	5	17	15	20	10	3	8
Quality controllers	5	15	24	18	14	8	6	5

SOURCE: Questionnaire Survey

per cent) of 'professionalized' laboratory technicians (i.e. those having a propensity to join professions) were already members of trade unions, as were 37 per cent of 'professionalized' draughtsmen and quality controllers; 18 per cent of planning and production engineers were also in unions.

Yet current members of professional bodies were rather more likely to have never been in a union than non-professionals, as Table 8.4 shows.

Table 8.4 Percentage of staff who have never been in a trade union

Occupational group	Professional	Non-professional
Draughtsmen	37	23
Laboratory staff	21	20
Quality controllers	50	40
Planning and production engineers	50	34

SOURCE: Questionnaire Survey

Furthermore, in all occupational groups there was a tendency among trade unionists to dismiss professional associations as irrelevant, and for their relevance to be more apparent to the non-unionist. This only became statistically significant in the drawing office ($x^2 = 1.377$ Df $= P < 0.005$). The figures are, however, not such as to suggest a dichotomy in attitudes which enables a separation into two distinctive groups of differently oriented technicians: 40 per cent of *non-union* quality controllers regarded professional bodies as irrelevant, as did a quarter of other technicians. Moreover, it seems that for at least a quarter of trade unionists (a third of draughtsmen) membership of a professional body would not be contradictory. To quote a laboratory technician who was a graduate member of the Institute of Electrical and Radio Engineers and Chairman of the local branch of the Association of Scientific Workers: 'The Institute is not interested in financial rewards: it is only interested in advancing theoretical knowledge.' It was, however, by no means certain that this would have been endorsed by all technicians. It became apparent that for most technicians the relevance of professional bodies was closely related to the ability of such institutions to obtain higher pay and prestige rewards for their members.

Socialization, education and training

It is often suggested that the offspring of manual workers are more likely to join trade unions than those coming from middle-class homes. The reasoning behind this hypothesis is that such employees will be socialized to accept a collective response to a personally-felt grievance rather than the highly individual one preferred by middle-class parents.[1] This kind of reasoning may be extended to the form of early socialization in the school environment experienced by the secondary-modern child as against the grammar-school product,[2] and so on through life; higher levels of achievement coinciding with adjustments or confirmation of what are 'appropriate' responses to the actions of those regarded as significant by the actor. In this way, the ethos of competition and individual achievement becomes reinforced and 'internalized'. Members of professional bodies are seen as relying on the prestigious recognition of a formal qualification which in itself is a mark of individual rather than collective achievement and therefore relates to a middle-class cultural environment in home and school.

(1) Family background

From the table below it can be seen that the majority of draughtsmen who were members of the union came from families in which the father was a manual worker. Laboratory assistants with manual workers as fathers also seemed slightly more likely to join unions, though neither among draughtsmen nor

[1] J. Klein, *Samples from English Culture* (Routledge, London 1965).
[2] S. Cotgrove, 'The relations between Work and Non-work among technicians', *Sociological Review*, Vol. 13, No. 2 (New Series) (July 1965).

Table 8.5 Current membership of a technicians' union by occupation of informant's father

Occupational group	Whether currently member of trade union	Occupation of father manual	Occupation of father non-manual	All
		percentage	percentage	
Laboratory staff	Yes	54	46	100
	No	50	50	100
Draughtsman	Yes	55	45	100
	No	42	58	100
Planning and production engineer	Yes	20	80	100
	No	45	55	100
Quality controller	Yes	40	60	100
	No	33	67	100

NOTE: SOURCE: Questionnaire Survey.

Categories of father's occupation:

Manual
1. Unskilled
2. Semi-skilled
3. Skilled

Non-manual
4. Clerical or routine
5. Upper-level inspector, supervisor
6. Managerial executive
7. Professionally qualified; in higher administration
8. Owned own business

laboratory assistants was the relationship statistically significant. On the other hand, union members among planning and production engineers were significantly more likely to come from white-collar homes than those of manual workers ($x^2 = 8.97$: $P < 0.005$).

It is evident that within these occupational categories the bulk of the unionists were from lower middle-class homes. This was not always so; for example, of 16 laboratory assistants who came from professional homes 12 were union members. There was also evidence of downward occupational mobility between generations among quality controllers and planning and production engineers. A dozen of the latter came from executive and professional homes and were currently union members.

In the group in which the desire for professionalization was most likely to lead to actual membership of an institution,

**Table 8.6 Current membership of a professional institution
by occupation of the informant's father**

Occupational group	Whether member of professional institution	Occupation of father manual percentage	Occupation of father non-manual percentage	Total
Laboratory staff	Yes	73	27	100
	No	57	43	100
Draughtsman	Yes	46	54	100
	No	60	40	100
Planning and production	Yes	37	63	100
engineers	No	51	49	100
Quality controller	Yes	33	67	100
	No	45	55	100

SOURCE: Questionnaire Survey.

that is, that of planning and production engineers, the influence
of a middle-class environment might be said to be evident though
not statistically significant. This was also true of quality controllers, especially those who placed their fathers in the executive/
professional categories. Only among laboratory assistants does
the relationship between early home background and professionalization work in the opposite direction; that is, technicians
from working-class homes were more likely to see professions as
relevant than those from the middle classes ($x^2 = 3.68 : P < 0.10$).

In a minority of firms, supervisors actively encouraged their
junior staff to work for professional qualifications. It seems
likely that generally the decision to join a professional institution
was seen as confirming the individual in his present status and
affirming his aspirations and expectations for the future. If,
however, this factor was taken into account it was apparent
that far from the 'will to achievement' being confined to
technicians from middle-class homes, it was evident that those
who were the sons of manual workers were rather *more* likely
to be extremely ambitious than those brought up in the homes
of white-collar workers.

(2) Education

If one supposes that union membership is inversely related to
the level of school attainment, then any trend towards educational dilution in recruitment to, say, the drawing office (see

Chapter 3) should work towards an increase in unionization—given that the union itself wishes to recruit less-educated members. Conversely, it might be argued that the raising of educational standards in the laboratories might work in favour of increased unionization there, since the raising of basic levels of entry to the higher (technologist) grades might create a whole new stratum of employees in which frustrated ambition could lead to some form of collective protest. The latter process might be said to be identifiable in the results of this survey, though it is by no means certain that the previous arguments hold good in any direct sense.

Even after taking into consideration their greater numbers, ex-grammar-school pupils entering the laboratories are significantly more likely to join a trade union than the laboratory technician who has attended another type of school. Of 222 laboratory assistants, only 46 had never been in a trade union, and of the 128 trade unionists, 126 were ex-grammar-school boys. The trend over a ten-year entrance period was towards an increasing number of grammar-school recruits for those unions organizing in laboratories. Over half (55 per cent) of laboratory assistants with A levels belonged to a union; and 62 per cent of those with O levels. By way of contrast, those (generally older) laboratory technicians with no formal school qualifications were only 37 per cent unionized. This type of correlation between union membership and school qualification was also present among quality controllers; 58 per cent of these technicians possessing A levels were in unions, a proportion which dropped to 39 per cent among O level G.C.E. holders.

On the other hand, former secondary-modern and secondary-technical pupils accepted the relevance of professional bodies to a greater degree than did ex-grammar-school pupils, especially among draughtsmen, where 44 per cent of the secondary-modern sample saw their relevance as against 34 per cent who did not. Against this may be placed the fact that the rate of unionization among draughtsmen with a secondary-modern education had not increased significantly over the ten-year period (i.e. 1958–1968). That is to say that although the sample showed a slight decline in the educational standard of entry to the drawing office, over that period the density of union membership in the drawing office had not increased as a result. By and large, the appeal of professional institutions for the low attainers among draughtsmen, quality controllers and planning and production engineers was somewhat higher than the attractions of unions, while high school-attainers among laboratory assistants were rather more likely to reject existing professional bodies than to accept their relevance. This obviously reflects the feeling of young men whose formal qualifications do not go beyond the sixth form.

The overall picture of the effect of home background and education is obviously not quite so clearly patterned as some sociologists have hitherto assumed it to be. It is only by a more detailed study of the nature of technicians' aspirations and the channels available to them through which they may achieve their ends *within* the firm that socialization and training become meaningful.

Other external characteristics that were tested included sex and marital status. Of 21 women in the sample, 12 were trade unionists (57 per cent), 8 of these being laboratory assistants. The majority tended to regard professional bodies as irrelevant in spite of their previously high level of school attainment. Among men, married technicians were at least twice as likely as single ones to join both trade unions and professional associations, a fact which becomes more significant when set against their comparative youth, low pay and high ambition.

Recruitment and training

The source of recruitment to the firm has very different implications for trade-union membership within each occupation. The significant influence among draughtsmen appears to occur among ex-apprentices who served their apprenticeship with their present employers: 68 per cent of these people had subsequently joined the draughtsmen's union. Fifty-nine per cent of those who had changed their firm during their career were not in current membership, though most had been members before joining their present firm. There was a high probability that if DATA was present in an office the new entrant from an apprenticeship would be recruited immediately, often with the co-operation of the A.E.U., to which union the employee had belonged whilst serving on the shop floor. Since most large firms have a higher proportion of apprentices than smaller ones, and since DATA has long-established recruitment and representational procedures within such firms, there is a strong chance that in any sample ex-apprentices will be in the Association while they are still at the firm at which they were trained. Thereafter the possibilities appear to decline as the level of commitment required for actually choosing to be a member upon entering a new situation gets higher.

The reverse was true of quality controllers, among whom externally-recruited men were more likely to belong to a technicians' union, though not usually until after joining their present firm. Planning and production engineers who had been promoted from the shop floor appeared to be significantly more likely to join staff unions than those from other sources of recruitment, and this provides us with a possible linking feature within these three occupational groups. In each of these cases potential unionists had been working in close contact with shop-floor workers either during their training or after.

Table 8.7   Current member of technician's union

Formerly member of manual workers' union	Laboratory staff		Draughtsmen		Quality controllers		Planning and production engineers	
	Yes	No	Yes	No	Yes	No	Yes	No
Yes	63	37	53	46	39	59	29	70
No	53	47	41	59	39	59	25	75

NOTE: Figures in percentages.   SOURCE: Questionnaire Survey.

From a comparison of Table 8.7 with Table 8.1 it appears that former experience in a manual union does have some socializing effect on laboratory assistants and draughtsmen, who are, of course, the most strongly unionized and who also appear to have most experience in manual unions. Quality controllers and planning and production engineers were much less likely to have had experience in manual unions than the former groups. This is largely explicable in terms of the 'bridging' nature of their previous occupation. Quality controllers came from a variety of backgrounds including H.M. Services; ex-draughtsmen among planning and production engineers had in fact often been members of DATA during their stay in the drawing office. Yet when an analysis was made of the careers of those who were previously in membership of a manual union, it was apparent that the break with manual unions most often took place through inter-firm mobility. When the previously skilled manual worker or supervisor is promoted, through his move to a 'quality controller' or 'planning and production engineer' post, he generally wishes to demonstrate this change in status. Not only that, but he believes himself to be much more exposed in his new position to the social and market pressures of management than are his colleagues in the laboratory and drawing office. The career of the man in a 'controller' or 'engineer' role takes him away from the external anonymity of the work-group. Yet because of his position in the socio-technical system he remains in geographical and physical proximity to shop-floor workers, and has a strong desire to demonstrate his social and market differentiation from them. For this reason, if he had changed his firm he was likely to have left the union in addition: 60–61 per cent of those who received promotion within the firm in which they did their training changed to a white-collar union after promotion.

It therefore becomes apparent that a too simplistic interpretation of Table 8.7 is misleading: the 'demonstration effect' of manual unionization did not operate in a straightforward manner. This becomes even more apparent when it is borne in mind that a large proportion (41 per cent) of laboratory assistants had no shop-floor experience and had therefore joined manual unions because these, and not technician associations, were recognized within their place of work as representing white-collar staff. Within at least four of the plants represented in the sample, members of the A.S.T.M.S. answering the question on their former membership of a manual union were referring to a union from which they had subsequently 'broken away' in an attempt to get a white-collar union recognized on their behalf. In other words, while membership of manual unions had provided a model for collective action, that action was directed at differentiating their goals from those of shop-floor workers rather than expressing a communality of interest.

It is often assumed that social or geographical proximity with manual workers leads automatically to an acceptance of similar forms of collective action. Experience as an apprentice is common to a high proportion of technicians in each group, and one might have expected a positive relationship between union membership and the number of apprentices within each group. This is true of ex-apprentices among draughtsmen and laboratory assistants, but the 'demonstration effect' appears to be a negative one in the case of quality controllers and planning and production engineers.

The belief that apprenticeship experience leads on to greater unionateness[1] in some direct manner is erroneous; it fails to appreciate the fact that the apprentices' work situation is viewed from a market position which is wholly different from that of most of his 'work-mates'. His membership or colleague group is comprised of other apprentices with similar high expectations and ambitions, often using an additional and different technical language from that used on the shop-floor. He is essentially a 'man of two worlds', and in many respects when he leaves the technical-college class for the workshop he does so insulated by a belief that his opportunities in life extend beyond those of the 'semi-skilled' operators who surround him in the work situation.

Unionization might therefore be an expression of a desire to maintain self-perceived status *vis-à-vis* manual workers in the face of persistent frustration, rather than a 'solidaristic' expression of unity with manual employees against management. The high turnover rate among members of the draughtsmen's union may be an indication of respondents' priorities, and their response to later questions seemed to confirm this hypothesis. For many the expansion of individual career opportunities was marked by a diminution in concern for group or collective status.

The effect of training on the development of professional attitudes has also been noted by most students of this subject.[2] One might have expected the long apprenticeship, averaging around five years, to have provided an occupational ethos for such groups as draughtsmen, which might have found expression in a desire for a professional institution. This was true of the 33 per cent of ex-apprentice draughtsmen who found professional bodies relevant, but the larger proportion, 46 per cent, rejected such a suggestion. Feelings were rather more evenly balanced among apprenticed laboratory staff: 27 per cent supported professional bodies, 32 per cent did not, and among the other two occupational groups 17–18 per cent of the samples found them relevant and a similar proportion irrelevant; most regarded present institutions with indifference.

[1] R. M. Blackburn, *Union Character and Social Class* (Batsford, London 1967).
[2] G. Millerson, *The Qualifying Associations*, (Routledge and Kegan Paul, London 1964).

I

Generally the greatest support for professional bodies came from among technicians trained in company-based staff training schemes: 65 per cent of the support for professional bodies among draughtsmen came from such ex-trainees. Among planning and production engineers and quality controllers 50 per cent of 'professionalized' technicians were ex-trainees. Laboratory staff were the exception to this rule; staff-trained technicians made up only 30 per cent of professionalized informants in the laboratory.

The large disparity which exists between common features of professionalization isolated by students of this phenomenon[1] and those displayed by our 'professionalized' technicians cannot be overlooked. Professionally aspiring employees have been described as attempting to gain institutional control of the training and entrance standards in their occupation; our informants were generally company-trained and most often lacked any external qualification. On the other hand, those who had achieved formal universally acceptable qualifications tended to reject existing professional bodies.

Formal qualifications and technicians' aspirations

At the time of this study the majority of the technicians in the sample population were heavily dependent upon the traditional means of induction and career progression. That is to say, the largest proportion was recruited and trained through either an apprenticeship scheme or shorter staff-training schemes. In both cases technicians were heavily reliant on part-time facilities for technical education or training. The sample population was relatively well qualified, yet it could hardly be described as formally highly qualified in any absolute sense; 34 per cent of the sample had no formal qualification whatsoever. It was evident that, for a sizeable proportion of those who *did* have a technical qualification, the part-time education provided by arrangement with the firm by the local technical college had been an important 'catching-up' device for former secondary-modern-school pupils. If employers had been forced to 'dilute' their standards of post-school entry (as some have suggested that they have), this did

---

[1] For a summary of the literature, see Archie Kleingartner, *Professionalism and Salaried Worker Organisation:* (Industrial Relations Research Institute, The University of Wisconsin, 1967).

not appear to have affected the aspirations of their recent recruits from secondary-modern schools, which remained as high as those with better qualifications.

Even after obtaining a minor qualification, the market position of the company-trained technician in relation to that of the graduate is poor and appears to be worsening. Despite the assurances of general and personnel managers that experience and 'ability to do the job' was given equal weighting with formal qualification, it was apparent from our survey data that technicians themselves were aware of a growing 'graduate barrier' blocking their access to the technologist career ladder.[1] In our interviews with middle management, supervisors and shop-stewards, there was a general consensus that no matter what was 'official policy', the recruitment of graduates as junior engineers or technologists was increasing very rapidly and was serving to close a promotion outlet. The belief in the availability of promotion outlets remained strong among management, largely because the present technical management and senior technologists were promotees. In many companies they were probably the last generation of this type of manager, though the stereotype remains a reality for them.

One of the major tenets of the 'class' explanation of trade unionism[2] is that of realization of limited chances through blocked promotion outlets: 68 per cent of laboratory staff with O.N.C. certificates or City and Guilds Certificates and 36 per cent with H.N.C. are members of trade unions. Among quality controllers the majority with City and Guilds Certificates (57 per cent) are in unions, as are 50 per cent of planning and production engineers with a similar qualification. The figures for union membership fall dramatically at higher levels of qualification, and this upper contour of qualification among union membership could therefore indicate inter-occupational differences at the point at which technicians adopt a collective view of their life chances: that is to say, the point at which they identify the myth that lies behind the managers' stereotype.

Most members of professional bodies were in possession of a

---

[1] There is also a tendency for this to hold good for careers in general management as well, but as yet the managerial career ladder remains the more open to the formerly lowly qualified employee, be he clerk or technician.

[2] To be distinguished from class-conscious action of the 'Klassen für sich' variety: see K. Marx, *Das Kapital*, (Foreign Languages Publishing House, Moscow).

higher external qualification.[1] Those who remained outside and merely aspired to this status may well have been frustrated by their own lack of qualifications. Of those who found professional bodies relevant, between 33 per cent and 40 per cent were totally unqualified, as already mentioned; taking the qualified technicians as a whole, they were much more likely to *reject* the relevance of professional bodies. Among well-qualified draughtsmen, the level of rejection was high: 45 per cent of H.N.C.s and half the laboratory technicians with H.N.C. either rejected them or were indifferent to them. But in the distribution of replies from those eligible to join professional associations there was a clear positive correlation with age; the older technicians with higher qualifications were much more strongly inclined to view professional bodies with favour. This was of course in marked contrast to the negative association between age and union membership which was observed in the laboratories and drawing office. Thus there does appear to be something of a 'generation gap' in the attitudes of members of these occupations to the respective merits of unions as against professional associations.

Rejection was highest among those with more marginal formal qualifications such as O.N.C. It might have reflected an over-adjustment to their inability to match the entrance qualifications of professional associations, a disability made worse by the recent closure of the major bodies by Royal Charter (see Chapter 3). The new status of those bodies affiliated to the Council of Engineering Institutions was, at the time of this survey, based on ultimate closure of membership to all except those in membership who were capable of passing a degree-level final examination. The need for some alternative to the recently founded Chartered Institutes of Engineering qualification which would make qualifications available to technicians was clearly evident in our survey findings. In answer to the question, 'Is there a need for a nationally recognized professional qualification for technicians?' over half of all informants across all of the occupational categories replied in the affirmative. Among planning and production engineers and quality controllers, technicians who were among the more 'professionalized' groups, over 60 per cent, were in support of some new kind of qualification.

---

[1] Our data cannot be precise here because of the tendency of members of any and all kinds of 'professional' body to claim graduate status as stemming from their status within that body.

Table 8.8 Educational qualifications of 'Unionist' technicians and 'Professional' technicians

Occupational group	Unionist/Professional	O.N.C.	H.N.C.	H.N.D.	C. and G.	FORCES	Degree/Dip. Tech.	None	Total
Laboratory staff	Trade unionists	(29) 24	(15) 12	(—) 0	(7) 6	(—) 0	(2) 1	(70) 57	(123) 100
	Professional bodies relevant	(9) 15	(16) 29	(—) 0	(5) 8	(—) 0	(6) 11	(21) 37	(57) 100
Draughtsmen	Trade unionists	(22) 30	(16) 21	(6) 8	(5) 7	(—) 0	(—) 0	(29) 37	(78) 100
	Professional bodies relevant	(17) 24	(25) 34	(—) 0	(10) 14	(—) 0	(2) 3	(16) 23	(70) 100
Quality controllers	Trade unionists	(6) 11	(1) 2	(—) 0	(16) 30	(2) 4	(1) 2	(28) 52	(54) 100
	Professional bodies relevant	(7) 18	(5) 13	(—) 0	(9) 22	(1) 2	(2) 4	(16) 40	(40) 100
Planning and production engineers	Trade unionists	(14) 15	(11) 12	(2) 2	(11) 12	(2) 2	(5) 5	(46) 50	(91) 100
	Professional bodies relevant	(3) 11	(6) 22	(—) 0	(5) 20	(1) 4	(3) 11	(9) 33	(27) 100

SOURCE: Questionnaire Survey.

NOTE: Figures in parentheses are absolute numbers.

Needs and aspirations

One very clear distinction made by many students of the process of 'professionalization' is between the 'job-centred' employee and one who is more instrumentally-oriented towards his work.[1] In this dichotomy, the indices used to measure the former as against the latter are the importance attached to intrinsic features of the work role and the task, as against extrinsic rewards for work.[2] Within the latter term are included market rewards such as pay, security and promotion outlets, but it also extends to describe satisfaction deriving from social relationships within the work-group and identity with the firm. Intrinsic rewards are usually taken to include satisfactions derived from the performance of the task itself and the use of personal skills and knowledge within the task environment.

If one accepts this hypothesis, one might have expected the professionalized workers to express the greatest strength of feeling towards intrinsic sources of satisfaction in their job, and unionists to express greater concern for extrinsic rewards.

The ordering of factors is in general very similar to that of the total sample (see Chapter 4). Both professionalized and unionized technicians give high place to an interest in their work; professionalized planning and production engineers and draughtsmen in unions score the highest levels of satisfaction in this respect. This comparison is probably a useful one to make, since both groups are in a relatively stronger market position than are the more lowly paid occupations, and therefore freer to discount extrinsic factors. Yet it seems possible to regard both unionized and professionalized technicians as 'job-centred' individuals for whom the level of satisfaction derived from utilization of skills assumes a particular salience in their whole pattern of needs gratification. Only a minority of unionized laboratory technicians gained much positive satisfaction from this source: quality controllers in unions also scored lower in this area than any other group, and satisfaction derived from their interest in work is much lower than that of other groups.

---

[1] See, for example: Louis H. Orzack, 'Work as a "Central Life Interest" of Professionals', *Social Problems*, 7 (Fall 1959), p. 127; John W. Riegl, *Collective Bargaining as Viewed by Unorganised Engineers and Scientists* (Bureau of Industrial Relations, University of Michigan, Ann Arbor 1959), p. 103.

[2] See J. Goldthorpe *et al.*, *Affluent Worker: Industrial Attitudes and Behaviour*, (Cambridge University Press, London 1970).

Table 8.9 Which of the following factors bring you satisfaction in your present job?

Occupational Group	Unionist/ Professional	People you work with	M.V.[1]	Utilization of skills	M.V.	Salary	M.V.	Fringe benefits	M.V.	Interest in work	M.V.	Identity with aims of company	M.V.
Laboratory staff	Trade unionist	68	0.53	62	1.59	48	1.30	54	1.31	76	0.89	29	1.05
	'Professionalized'	80	0.93	64	1.51	44	1.00	54	1.39	78	0.99	29	1.57
Draughtsmen	Trade unionist	70	0.66	70	1.29	54	1.05	54	1.34	80	1.05	31	1.45
	'Professionalized'	78	0.74	76	1.41	56	1.31	56	1.32	70	0.75	29	1.31
Quality controllers	Trade unionist	70	0.84	44	1.87	56	1.51	60	1.67	34	0.85	60	1.82
	'Professionalized'	80	0.86	68	1.15	82	1.23	44	1.91	76	0.59	74	1.92
Planning and production engineers	Trade unionist	70	0.81	70	1.32	64	1.40	48	1.18	78	1.01	69	0.65
	'Professionalized'	74	0.83	66	1.30	88	1.16	56	1.82	80	0.89	54	1.33

NOTE: (1) M.V. = Mean Variance.

SOURCE: Questionnaire Survey.

When the sample were asked to give a priority to changes they most desired managers to make, both laboratory and quality-control technicians gave a high priority to these factors, but union members scored significantly higher than non-unionists. When related together their replies suggest that quality controllers who join unions experience a very much higher degree of frustration arising out of the non-utilization of their skills than technicians in any other category.

The importance of work-group relationships is emphasized by these results, and provides a vivid contrast with the pattern of feelings expressed towards the company. When the attitudes of either unionists or would-be professionals in the laboratories and drawing offices are compared with those of quality controllers or planning and production engineers, their company loyalty is very significantly lower, whilst their feelings towards their work-group are at least as warm. It appears that the more-important element in determining the attitudes of technicians who are divorced from line activity and who are relatively isolated as a group within the social and work structure of the firm may be the work-group itself rather than the firm as a whole. When provided with an external occupational reference group such as draughtsmen possess through their occupational identity across several industries, the potency of the division of interest within the firm becomes even more significant in its implications for industrial relations. By contrast, professionalized quality controllers express the most positive feelings of company loyalty. Like planning and production engineers, they are also more likely to derive satisfaction from their present level of salary, and, significantly, from fringe benefits. It may also be recalled that dissatisfied line technicians were more likely to regard changes in fringe benefits as being important.

The pattern of needs and satisfactions which emerged from the questionnaire survey supports the view formed in interviews that some technicians were much more likely to regard the company as a major source of needs gratification, especially in terms of their expected career, whilst others were likely to see it in a much more transitory and instrumental fashion.[1] The tech-

[1] See J. V. Eason and R. J. Loveridge, 'Management Policy and Unionisation among Technicians', in *Proceedings of an S.S.R.C. Conference on Social Stratification and Industrial Relations* (Social Science Research Council Cambridge 1969), for a more detailed exposition of differences between 'locals' and 'cosmopolitans' among technicians.

nician's subjective view of his market situation bore little relationship to his actual pattern of job movement between firms, or even to his expressed intention to move (see Chapter 7). His objective chances of either geographical or social mobility were to an extent a function of age, home ownership, and several other situational factors which acted as constraints upon movement, though not upon aspirations. The higher age of quality controllers and planning and production engineers meant that they were most likely to have made several moves before arriving at their present position.

Yet, as is shown in Table 8.10, most were not yet prepared to admit to the growing limitations upon their opportunities for either promotion or job movement. *Those with the longest-lasting aspirations were the highest school-attainers or apprenticed technicians,* though often the later qualifications of these people were minimal in formal terms. Hence it was possible for many technicians to see their career chances as being related to the long term well-being of a single firm at one level of cognition, whilst at a different level experiencing a high degree of dissatisfaction with many features of their work, especially intrinsic factors, such as the non-utilization of their skills, which led them to declare their intention to leave the firm's employ.

Table 8.10   Current trade union members who were
'not really ambitious for promotion'

Occupational group	Unionist or non-unionist	True	False	Don't Know	Total
Draughtsmen	Trade unionist	12	77	11	100
	Non-unionist	11	80	9	100
Laboratory staff	Trade unionist	12	84	4	100
	Non-unionist	15	77	8	100
Quality controllers	Trade unionist	22	72	6	100
	Non-unionist	20	74	6	100
Planning and production engineers	Trade unionist	40	55	5	100
	Non-unionist	2	96	2	100

SOURCE: Questionnaire Survey.

In the event it was difficult to discover whether many in fact did so. But it seems that for the low school-attainers or staff-trained men mobility had been relatively low in the past. For such men the desire to differentiate themselves from shop-floor workers and to establish some degree of autonomy for themselves within the line environment became expressed in a desire for professional membership. The relatively foot-loose and more qualified draughtsmen were more likely to be found in unions. For the latter, as for laboratory assistants, the work-group within the drawing office or laboratory appeared to be a much more important referent in deciding what were appropriate attitudes than it was for line technicians. Of the 47 per cent of draughtsmen and 36 per cent of laboratory technicians who intended to leave their present firm for 'better prospects' the large majority of these were in unions. They were also young and relatively well qualified.

If a *lack* of ambition were widespread among technicians it would provide strong evidence for the belief that unionization is associated with a solidaristic form of group consciousness in which an increase in status is to be seen in terms of group escalation. This was the case among the 40 per cent of unionists among planning and production engineers shown above; these were generally older men in positions which they obviously regarded as terminal. Of the minority of all technicians who were unambitious, the majority were attracted neither to unions nor to professional associations. The exceptions were draughtsmen, among whom the unambitious were significantly more likely to be 'professionalized', but, as in other groups, not likely to be unionists. As suggested above, these professionalized draughtsmen were most likely to be staff-trained men. On the other hand laboratory assistants in trade unions were actually the *most* ambitious of technicians. The data seem to suggest, therefore, that while a growing realization of career-blocked prospects is a significant factor in the propensity to join unions, it does not necessarily lead to the abandonment of individual career aspirations.

Trade unionists were everywhere more likely to feel 'blocked' in their ambitions than not, and this was especially true in the laboratories and drawing offices. The table below demonstrates the strength of association between 'blocked promotion' and membership of a trade union among laboratory staff; this contrasts with planning and production engineers, among whom frustrated ambition does not appear so likely to lead on to unionization. Clearly there are differences in the degree to which frustration leads on to union membership. One possible reason might lie in the degree to which the barriers to promotion seem immutable and impermeable. When asked what reasons for promotion were believed to be currently in use within the firm, 53 per cent of the laboratory assistants who were in unions opted for technical qualifications. Yet laboratory assistants were also significantly more likely to see 'length of service' as a currently used criterion of promotion and if they were unionists than if they were not, and to see 'drive and initiative as going unrewarded'. These apparently contradictory attitudes reflect ambiguities in their work situation in which current superordinates appear to have arrived in their present position by dint of long experience, while new appointments reflect the more recent criterion of formal qualification. In neither case do the immediate efforts to achieve their ambitions made by young and aspiring technicians appear to stand comparison with the achievements of others who have been successful in the competition for promotion in the past.

Table 8.11 Technicians who were keen to be promoted but also believed there to be insufficient opportunities, shown by current union membership

Occupational group	Trade unionist or non unionist	You are keen to be promoted but there are insufficient opportunities		
		True	False	Don't know
Laboratory staff	Trade unionist	57	49	2
	Non unionist	41	47	4
Draughtsmen	Trade unionist	44	44	7
	Non unionist	49	51	5
Quality controllers	Trade unionist	41	35	7
	Non unionist	52	61	4
Planning and production engineers	Trade unionist	19	24	11
	non unionist	70	41	35

NOTE: Figures in percentages.   SOURCE: Questionnaire Survey.

Quality controllers were significantly more likely to see 'drive and initiative' as going unrewarded, and their qualifications as being unrecognized by management, if they were in unions. On the other hand there was no evidence for believing that unionists felt that their union affiliation made it harder for them to get promotion.

The generally ambitious, generally young trade unionist seemed more likely to resolve the apparent contradictions in management promotion policy by describing 'favouritism' in selection the reason for his failure. Among the older and more senior planning and production engineers this was truer of non-unionists (59 per cent) than of unionists (8 per cent), again suggesting that the level of dissatisfaction, and the means chosen by technicians to resolve a high degree of cognitive dissonance created by their career environment, should be analysed in a separated manner. In other words, dissatisfaction at blocked aspirations did not automatically lead to union membership, and it was certainly not true that union membership was associated with any marked degree of 'solidaristic' class identity.[1] Over 60 per cent of all technicians, whether unionized or not, felt promotion came 'from being in the right place at the right time'. There seemed every reason to believe that unionists wanted to be in that place at least as much, and possibly rather more, than non-unionists!

Only among draughtsmen was there any marked recognition of a company career-development policy which met the traditionally accepted standards of professional employees; 49 per cent of professionalized draughtsmen felt 'drive and initiative' *were* rewarded, 44 per cent felt merit was, and 43 per cent technical qualifications. Again it is necessary to point out the duality in meaning attached to the relevance of professional bodies. Most of these 'professionalized' draughtsmen were staff-trained men with highly specific skills. Their satisfaction with rewards for 'merit' was related to the development of 'graded' salary scales within the drawing office, to which some differentiation in job content, work authority and—most important—differentiated job-titles had been attached. More outward-looking youngsters in laboratories rarely had such multi-grade promotion ladders (the gulf between graduate and non-graduate was becoming increasingly obvious), and indeed, like ex-apprentices in other departments, would probably be less willing to accept them as representing a career-ladder—at any rate until they reached their thirties.

Despite the multi-variate nature of their longer-term concern

[1] See, for example, H. H. Hyman, 'The Value Systems of Different Classes', in R. Bendix and S. M. Lipset (eds.), *Class, Status and Power* 2nd edition (Routledge and Kegan Paul, London 1967).

over personal and group status, most of the technicians' worries were subsumed beneath their more immediate need for a higher salary. This was by far the most significant of the changes required of management in order to bring satisfaction and status in their job. This was particularly true of laboratory technicians. For unionists in laboratories current dissatisfaction was more than twice that of their professionalized colleagues. Whatever the frustrations of technicians in respect of their task interests, they were generally far readier to express dissatisfaction with their salaries, whether professionalized or trade unionists. It is evident that laboratory assistants are, as one might expect of the youngest category in our population, the least well paid. Yet draughtsmen of a similar age are better paid, while the span of earnings among quality controllers and planning and production engineers is as much a reflection of the varying statuses that these titles cover within this survey as age *per se,* or the local market position. So that as a group, laboratory technicians were least well off in real terms, and this cannot be dissociated from their much greater propensity to unionization.

In general technicians who rejected professional bodies were most likely to be in the middle income groups, and across three of the four occupational groupings the effect of professionalization appeared to decline thereafter with increasing levels of salary. Among laboratory assistants, for example, those earning under £900 were almost equally divided between those who could, and those who could not, see the relevance of professional bodies, but thereafter professionalization steadily declined in effect as salary increased. The exception to this rule was to be found among draughtsmen, of whom the largest proportion (23 per cent) were earning over £1,600 a year. This pattern is one that would support our earlier hypothesis that professional associations have an especial appeal to those who feel their position to be most socially 'marginal'.

The perception of a professional association which these technicians possess is that of a body primarily designed to clarify and define their status, rather than to reinforce an existing occupational identity. The more highly-paid laboratory technicians included those who had already achieved considerable advancement and saw little purpose in professional bodies. Senior draughtsmen were aware that to progress they had to move out of the drawing office. They were very much more likely than other groups to choose the title of 'engineer' (usually 'design engineer') to describe their role; perhaps this was a mark of their desire to progress beyond their present identity within the drawing office.

It was only in the laboratories that there was an inverse correlation between salary level of technicians and their propensity to join unions. Seventy-five per cent of laboratory assistants earning under £900 were unionized, as against 12 per cent who were not, and those lowly paid union members made up over a third of all unionized technicians in the laboratories. Elsewhere the modal salary level of unionists tended to move upwards within groups, so that the highest numbers among quality controllers were earning £1,000 to £1,099, and the more highly paid unionists were to be found among draughtsmen and planning and production engineers (£1,200 to £1,290). The *absolute* income level was not therefore of general importance to union membership outside the laboratories. One must ask whether *relative* income levels were important, and if so, relative to what or to whom?

When confronted with a list of occupational pay-groups, informants generally chose that of similar status to their own as the group with whom they compared themselves. A cross reference with other indices such as age, qualification, and method of recruitment, showed that technicians promoted from the shop floor within the same company structure were most likely to be using manual workers as their reference group for comparing earnings and possibly as an index of the waning status of technicians. Furthermore, when asked whether the pay of technicians was too low in relation to shop-floor earnings, 92 per cent of trade unionists among laboratory technicians, 87 per cent among draughtsmen, 93 per cent among quality controllers and 70 per cent of union members among planning and production engineers answered 'yes' or 'definitely yes' to this question.

The degree of saliency attached to shop-floor earnings as a reference point came through even more markedly in interview data and in the embittered comments written into questionnaire answers. Yet the evident feelings of relative deprivation expressed in both the weight and degree of opinion on differentials in these replies does *not* inevitably lead on to union membership. The *aggregate* numbers of discontented draughtsmen and quality controllers were greater among non-unionists than unionists. Indeed, such discontent was almost as likely to be associated with positive feelings towards professional bodies as towards unions (i.e., there was no significant difference between them).

Formal collectivization of either form appeared to have little overall effect on technicians' attitudes to what *should* be the most important determinants of company policy on technicians' salaries. If anything, criteria based on individual achievement seemed accentuated in the opinion of union members. When offered three choices from five alternatives—individual skills, formal technical qualifications, amount of responsibility, seniority, traditional differentials—together with an open-ended option, most selected responsibility above all else. But 81 per cent of the draughtsmen in union membership chose individual skills first, and this factor also received the support of 75 per cent of the planning and production engineers and laboratory assistants and 72 per cent of the quality controllers. In every case except the last, unionists were more likely to choose this factor than non-unionists.

Opinion among 'professionalized' technicians was not significantly different, but their emphasis on technical qualifications shifted upwards to second place. This was particularly true of the technically well qualified but poorly paid quality controllers. Many of these 'professionalized' employees were categorized as weekly paid staff, and, as is known from interview data, were extremely sensitive to the low esteem in which their qualifications (for example the City and Guilds Certificate) were held by management. Thus 68 per cent 'professionalized' quality controllers gave first priority to the worth of technical qualifications, as against 37 per cent of unionized laboratory assistants. Perhaps this was for the rather obvious reason that the latter were much less likely to have gained a *technical* qualification, or any other, since they had left the sixth form (i.e. since A levels), and this fact also effectively excluded them from any quasi-professional body as well as from promotion.

All groups showed a large majority (between 76 per cent and 96 per cent) in favour of schemes offering differential rewards based upon individual performances. This majority did not appear to be significantly influenced by the fact of union membership, nor even 'professionalization', except among laboratory assistants. All this seems to support the view that technicians may tend to join unions—and perhaps also professional bodies —for purely calculative purposes and may regard either type of body as being instrumental to the achievement of their individual aspirations rather than as ends in themselves.[1]

This view was borne out in Chapter 6 (on conditions of employment), where the remarkable degree to which all groups of technicians, but most particularly laboratory assistants, refer to shop-floor workers was emphasized as being one of the major factors in bringing about increases in technician earnings. The majority of laboratory technicians, and between 38 per cent and 41 per cent of all other groups, regarded shop-floor workers as their primary source of 'leverage' in individual or collective negotiations over earnings. The bargaining skills of unions rated a poor second, even among union members, except in the drawing office. Not only did DATA members unanimously regard their union as the most important influence in bringing about salary increases, but so did a large proportion of non-members' which gave the union an overall majority of supporters in the drawing office.

Once again the *fact* of union membership seems to be divorced from the feelings of respondents. Hence many laboratory assistants in unions attached more importance to the actions of shop-floor workers than to the bargaining power of the union which they had joined. Many draughtsmen who were *not* in a union nevertheless regarded it as being important in determining the wider market position of their occupation. The former were often working in quality-control functions which brought them into contact with manual workers; their awareness of manual unions may therefore have been greater as a result.

The large majority of those who placed management decisions as the most important among the factors in determining their salary were 'professionalized' workers, but for all quality controllers, as for planning and production engineers, considerable importance was attached to 'management deciding what you are worth'. This ordering of priorities appeared to support the hypothesis that these two groups seemed to perceive themselves as having a greater degree of dependence upon a localized demand for their skills and upon a more personal form of bargaining: 47 per cent of planning and production engineers saw salary increases as following from 'how forcefully you can put your own case', a fact which obviously relates to their personal position within the local management hierarchy.

[1] J. Goldthorpe *et al*, op. cit.

Despite the high degree of market vulnerability of these 'locals' or company-oriented line technicians, they were generally those likely to wish to resolve their feelings of market 'isolation' and insecurity through the medium of professional associations rather than unions.

Management as a normative reference group

The primary focus of technicians' ambitions is demonstrated in Table 8.12 below. It seems that only a minority would reject a management position in a manner which showed a positive dislike for the suggestion, and these tended to be older technicians. For the largest proportion of each occupational group, promotion to management appeared as an acceptable goal. The precise meaning of the suggestion would, of course, vary according to the informants' position in the firm and the nature of the firm,[1] yet the focus of ambition for young laboratory and drawing-office technicians is quite clear.

Among professionalized quality controllers and planning and production engineers the desire to become a manager was even more marked (68–72 per cent). This may be regarded as a point of differentiation between some of these company-oriented line technicians. That is to say, union members are significantly more likely to reject managerial promotion when it is suggested to them, while professionals grasp the propostion much more willingly. These were people who had 'adjusted' to their market position in other ways as well.

Table 8.12   'Would you like to be a manager one day?' by union membership

Occupational group	Unionist or non-unionist	Would like it	Not bothered	Would dislike it	Don't know
Laboratory staff	Trade unionist	60	15	20	5
	Non-unionist	62	18	16	4
Draughtsmen	Trade unionist	50	20	26	4
	Non-unionist	52	18	16	4
Quality controllers	Trade unionist	46	17	31	6
	Non-unionist	54	15	28	3
Planning and production engineers	Trade unionist	45	15	35	5
	Non-unionist	76	5	11	8

NOTE: Figures in percentages.   SOURCE: Questionnaire Survey.

[1] For a discussion of the implications of these differences see Chapter 5.

The impression of the lack of any wide measure of a collective identification of interest with other workers appears to be confirmed in Table 5.4, in which there is a uniformly high level of identification—not with shop-floor workers, but with management, difference of degree only existing between planning and production engineers (as the most closely identified group) and the 51 per cent of laboratory staff who identified with management. It did not therefore appear that estrangement from management was either a necessary or sufficient condition for union or professional membership. With the exception of quality controllers, all occupational groups of unionists shared an over-all majority wish to be identified with a managerial outlook. Across all occupations only 15 per cent or fewer of the union sample populations found their outlook to be more compatible with shop-floor workers than management. Quality controllers were exceptional in more ways than one. While 46 per cent of unionists identified with management, 17 per cent of *non-unionists* felt their outlook to have more in common with the shop floor. Once more 'the fact of union membership is less important than the motive for joining.'[1] For the exposed and company-dependent quality controller the act of joining a union may appear a hazardous undertaking, despite his ideological commitment in that direction.

Nor were other differences necessarily related to membership of an external organization. In the widespread identification with the managerial outlook, the offspring of manual workers seem to identify with management to a somewhat greater degree than those whose early socialization was that of a middle-class home. Later training also seemed to have a contrary effect to what might have been expected in any simplistic application of a class theory. For example, draughtsmen who have been trained in apprenticeship schemes are more likely than not to strongly identify with management, as well as to join unions!

Management response

This is a subject for a separate chapter, but it is clear that in the process of unionization or professionalization the management response to the expressed need of technicians is at least

[1] D. Silberman, 'Clerical Ideologies – a research note', *Sociology*, June 1968.

as important as the initial impetus in determining the structure
and course of subsequent events. This response may be seen as
having two main strands: firstly, in terms of the internal handling of the man-management function, sometimes referred to as
human relations; secondly in terms of the external relationships
of the firm with unions and professional associations. The
two may be closely interwoven in the warp and weft of work
interaction unless some attempt has been made to keep them
formally exclusive. It was, however, extremely difficult for the
formalized labour-management policy to be pursued in a manner
which did not reflect at some point the views of the two parties
at a more fundamental social and inter-personal level.

When asked whether management took a reasonable interest in
technicians there were almost twice as many in all groups who
were ready to make negative rather than positive replies. Feeling became even stronger over the question of their acceptance
into the management team, and written-in comments contained
a great deal of bitterness. This was carried over into opinions on

Table 8.13   Would you agree that you are part of the
management team?

Occupational group	Very strongly agree	Strongly agree	Not sure	Strongly disagree	Very strongly disagree	Don't know	All
Draughtsmen	1	12	19	31	26	10	100
Laboratory staff	4	11	20	31	29	3	100
Quality controllers	5	9	18	24	35	7	100
Planning and production engineer	7	22	20	28	16	7	100

NOTE: Figures in percentages.   SOURCE: Questionnaire Survey.

group status, in which the sense of collective deprivation is
clearly apparent (Table 5.7). The relationship of these feelings to
union membership and to professionalization is by no means a
straightforward one. For example, laboratory assistants who
suffer feelings of rejection and status deprivation are significantly more likely to be in unions than not, and vice versa.
Among planning and production engineers the most dissatisfied
people were those outside the unions. For these technicians, as
for the quality controllers, the attractions of professional associations for people who felt their status to be low and who were

'definitely not part of the management team' were very great. What is perhaps even more remarkable is the fact that nearly all technicians who regarded themselves as already part of the management team dismissed professional bodies as irrelevant. This suggests that the hypothesis sometimes advanced that a necessary condition for 'professionalization' may be an affinity to management,[1] might be modified to take in a growing feeling of separateness—albeit an unwilling self-exile from the internal sources of legitimacy within the firm. Thus the relevant first step to collectivization for the planning and production engineer or other technician in a similar work-situation could be seen as taking a professional form.

What are the conditions which these deprived groups of technicians see as being conducive to their joining a union and/or a professional association? Arising out of this question comes a second: within the sample population, all of whom appear to experience high levels of deprivation (though the pattern of their needs could be different), why should there be *differences* in the propensity to join unions and/or professional associations?

It has been shown that the technicians covered by this survey tended to identify with management. It has also been demonstrated that technicians were overwhelmingly career-oriented, and that although this provided the principal source of their frustrations, it also tended to increase the extent of their dependence upon management, especially among line technicians. Even the most geographically mobile technician would place a higher value on a 'clean copy-book' than a less career-oriented manual worker. Thirdly, most unionists appeared to display an 'instrumental' acceptance of union membership rather than an ideological one. Not only were they generally more concerned with using their bargaining strength to improve or restore their earnings differential above that of shop-floor workers, but they also displayed remarkable support for any reward schemes based on individual merit. These three factors have been put forward as major influences structuring the historical relationship between management recognition of a union and its subsequent growth.[2] The

[1] K. Prandy, *Professional Employees, A Study of Scientists and Engineers* (Faber and Faber, London 1965).

[2] G. S. Bain, *Trade Union Growth and Recognition*, Research Paper 6, Royal Commission on Trades Unions and Employers' Associations, (H.M.S.O., London 1967).

findings of this survey tend to modify the thesis that success in union recruitment follows from management recognition and approval.

Union recognition

Unions representing technicians were or had been recognized by eleven of the fourteen plant managements co-operating in this study. This was not always extended to all categories of technicians. Draughtsmen were the group most universally represented by a union—the Draughtsmen's and Allied Technicians' Association. ASSET or the A.Sc.W. had gained full recognition in only two of the plants, though it had pockets of membership in every plant visited in the study. In the two steel plants the A.S.T.M.S. was in direct competition for members with the British Iron Steel and Kindred Trades Association (BISAKTA), the largely manual workers' union organizing in the steel industry. Similarly, some manifest competition, accompanied by dual membership and in more extreme cases by threats of A.S.T.M.S.-led strikes, had occurred in several companies representing a complete span of industrial and technological contexts. Usually this conflict occurred between the Amalgamated Engineering Union (supervisory branch) and the A.S.T.M.S., but it also involved more unusual organizers of technicians such as the Union of Shop and Distributive and Allied Workers, or the T.G.W.U.

The dichotomy between the threat of a strike and the deed itself was revealed in the opinions recorded during interviews. The study produced evidence of the current changes in the actions of formerly 'conservative' groups of white-collar employees. One such group of laboratory assistants engaged in testing samples in a steel strip-mill overcame any inhibitions imposed by constraints of group status or personal promotion prospects and withdrew their labour during the course of a recognition dispute while our study was proceeding within the organization.[1] The resultant change in attitudes is not shown in our tables because data from the pilot study has not been incorporated in the statistical tables contained in the report. However, it is clear from the subsequent actions of these technicians that their outlook has changed considerably, and that collective action has taken on a more 'strategically' militant form as a result of

[1] See J. V. Eason and R. Loveridge, op. cit.

their initial experience of strike action in support of the recognition of the A.S.T.M.S. rather than BISAKTA.[1]

On the other hand, in one engineering firm management had refused to recognize ASSET a few years ago, after an outburst among technicians and supervisors over a perceived 'unfairness' in salary increases. Subsequently the grievance was righted, but management remained firm in their refusal to recognize the union. Interest in the issue of recognition faded among technician employees outside the drawing office. The membership of DATA included all draughtsmen at that time, but in subsequent years it fell off to a negligible proportion despite the continuance of formal recognition for their union. In other plants where ASSET or the A.Sc.W. *had* been recognized, either fully or partially (see Chapter 3), membership often dropped dramatically following formal recognition.

The reason for this was twofold. From the employees' side, the level of leadership and personal servicing offered to members was vital when they clearly had only a minimal interest in the long-run well-being of the union *per se*. From the management the level of commitment or antipathy shown by them towards technician unions, whether formally recognized or not, created the environment in which membership attitudes were formed.

Union members co-operating in this survey were generally more likely to see management attitudes as being favourable to staff unionism. The important exception to this was, as can be seen below, in the laboratory. They were, it will be remembered, also the group with the highest feelings of rejection and status deprivation; it was apparent that for many managements the act of joining a union in the laboratory was quite different from joining the union in the drawing office. While it had been accepted that draughtsmen had an organization of their own which had gained a legitimate and accepted place within national grievance procedures, within the same plant, it would also be held to be quite wrong, and indeed a mark of disloyalty to management, to be in another technicians' union if the employee were a laboratory or line technician. Possibly these technicians were perceived as being socially 'closer' to management.

---

[1] See L. R. Sayles, *Behaviour of Industrial Work Groups*, (Wiley, New York 1958), p. 5 ff, for a theoretical explanation of these terms.

Table 8.14 **Perceived attitude of management to staff unionism, by trade union membership**

Management attitude	Draughtsmen		Laboratory staff		Quality controllers		Planning and production engineers	
	Trade unionist	Non-unionist	Trade unionist	Non-unionist	Trade unionist	Non-unionist	Trade unionist	Non-unionist
Favourable	40	24	27	20	30	19	25	21
Neutral	33	40	25	30	41	30	55	23
Unfavourable	21	20	41	27	21	18	15	40
Don't know	5	13	7	20	7	31	5	17
Total	100	100	100	100	100	100	100	100

SOURCE: Questionnaire Survey

The fact that union members were often rather more inclined to see management as favourably disposed towards their unionism may have stemmed from their experience in testing management feeling. It is hardly surprising that planning and production engineers, who experience such high levels of deprivation, were also less inclined to join unions, given that they were likely to regard their superordinates as being opposed to such affiliations. The initiation of union recognition procedures was rarely one experienced by informants in the survey.

The importance of the card-carrying member (and to a larger extent the professional member) who acts as a 'pioneer' in introducing his organization into a plant cannot be underestimated. That quality controllers and planning and production engineers were the least organized may be due to a lack of sufficiently motivated 'pioneers' rather than to a positive concern for management feelings. The differing work situations of the respective occupational groups of technicians had an important effect upon their ability to create and to *maintain* a union 'cell' of members. For technicians working in a socially exposed work situation, rather than in a cohesive and geographically compact workgroup, the significance of their union membership was discovered in their inter-personal relationships with their immediate superiors on the one hand, and with their subordinates on the other. This was obviously less true for laboratory assistants and draughtsmen than for quality controllers and for planning and production engineers.

Of more importance to the draughtsmen's union as well as to unions recruiting younger quality controllers, were the effects of labour turnover upon membership within any given firm. This was found to be particularly evident in the plants in the South-East and North Midlands, where labour bottlenecks in the external market led to a constant flow of qualified young draughtsmen. Many of these men lacked a commitment to their union of a 'moral' or ideological kind which would have led them to maintain membership over a number of multi-various places of employment and differing employer responses. Clearly a very positive indication by management that they *expected* their technicians to belong to a union would have transformed the nature of membership. Instead of carrying membership from job to job it would become perceived as a condition of advancement, or at least an aspect of normative behaviour within the immediate work situation. This response was unusual even in the drawing office, where it was more normal for Chief Draughtsmen to recall days when DATA really *was* more like a professional association (an almost universal response among these members of middle management).

The importance of employer recognition to the growth of technicians' unions has been brought out by the study of national statistics undertaken by Professor G. S. Bain.[1] Yet the *decay* of union organization among technicians in the plants we examined was as widespread as the growth. While not wishing to detract from the importance of Professor Bain's contribution to a macro-level theory of institutional growth, it is important to recognize that it subsumes much that is fundamental in organizational and inter-personal behaviour at plant level. 'Decay' often takes place because of lack of leadership and membership commitment *after* long-standing management recognition for a union within the plant or sector from which it recruits. This criticism may be extended to the second major factor cited by Professor Bain as being determinate in the growth and density of white-collar unions: that of the concentration of white-collar employment. Though it was clear that where there were a large number of technicians within a plant the possibility of new leaders arising to keep the union going was greater, in fact the continued dependence of ever greater numbers of members upon the same small group of 'pioneers' appeared to be a much more usual phenomenon. This usually implied that the

---

[1] In G. S. Bain, *The Growth of White Collar Unionism* (Oxford University Press, 1970).

standard of servicing declined and members became progressively more likely to lose interest.

The 'check-off' system, by which the employer deducts union subscription at source, was used at only two plants. Elsewhere it was not unusual for an employer to continue to meet a union representative knowing that he represented an ever smaller proportion of his work-force.

Homogenization and communications

The supposed characteristics of bureaucratic organization have been described at great length by many theorists. The division of labour may lead to boring and repetitive tasks for an increasing number of workers. This has been seen to be true of the roles performed by laboratory and quality-control technicians. In addition, expanding organizational size may lead to centralized decisions being passed along extended lines of communication. Substantive and procedural rules become increasingly standardized and the individual personality is subsumed beneath a homogeneous mass: a digit in a grading system.

The technicians' view of intra-firm communications has already been examined at some length in Chapter 5. Occupational differences became clearer in the pattern of association between dissatisfaction and unionization or 'professionalization'. Among laboratory assistants and planning and production engineers our informants in trade unions were most likely to be satisfied with communications downwards on technical matters (36 per cent–38 per cent). Draughtsmen were slightly less so, and quality controllers very much less so (15 per cent). This latter group of unionists was unique in the strength of its condemnation of management's ability to communicate on technical matters. It was suggested in Chapter 5 that many of their problems arose out of the multiplicity of roles and the concomitant allegiances and responsibilities which were developed in the work situation of the line technicians.

On the other hand, there was a wider and stronger association between the views of all technicians on promotion procedures and their propensity for 'professionalization'. In three of the occupational groups the professionalized technician was significantly more likely to regard communications on promotion procedures as good or very good than were those who rejected professionalism.

Table 8.15 Technicians' view of communications on promotion

Attitude	Laboratory staff		Draughtsmen		Quality controllers		Planning and production engineers	
	Profes-sional	Non-Profes-sional	Profes-sional	Non-Profes-sional	Profes-sional	Non-Profes-sional	Profes-sional	Non-Profes-sional
Very good/good	47	9	49	(0)	16	11	4	14
Average	36	39	43	37	27	28	35	38
Very bad/bad	17	45	8	54	47	33	61	47
Total	100	100	100	100	100	100	100	100

NOTE: 'Don't knows' included in total of 100.　　　　SOURCE: Questionnaire Survey

For an explanation it should be remembered that professional-ization was quite strongly associated with staff-training schemes in the drawing office, and with lack of ambition. This became progressively less true as one moved across the columns in Table 8.15; the 'professionalized' planning and production engineers showed the highest degree of doubt about the system of promotion. These were people with high formal qualifications whose immediate future seemed to be particularly difficult to predict.

Upon promotion, and more particularly on changes in product which might affect their jobs, as well as upon communications relating to their department's prospects, many technicians ex-pressed disquiet. A preliminary analysis of the data suggests that faulty communications in the larger multi-plant companies tended to generate most dissatisfaction in this respect. The fact that four of such plants had densities of union membership ranging between 60 per cent and 76 per cent might lead one to suspect that bad communications on market issues might be a significant element in the process of unionization. Yet a plant attached to the second-largest international combine included within the study (and one of the largest in the world) had a density of union membership amounting to only 2.5 per cent. Also one might compare the 65 per cent unionization at one plant, belonging to the largest of the companies in the study, to the 38 per cent at another belonging to the same company which is run along similar, centrally agreed principles. It did not appear that size or geographical separateness of themselves led to a desire to join a union, but rather the amount of 'threat' which appeared to be contained within this growing isolation from decision-taking in the product and labour market.

The factor that generated the most generally expressed unease in interviews was that of redundancy stemming from 'rationalization', which had accompanied the growth of these multi-plant firms. In at least five of the fourteen plants included in the survey, major restructuring of the company's employment policy was in progress during the course of the survey, in which 'rationalization' was the aspect of management's goals best known among our informants. It should not, however, be thought that this provided an unusual bias to our sample. In some companies this was the second or third of such exercises to be carried out in recent years.

The second dimension in dissatisfaction was that mentioned in Chapter 5 as existing among planning and production engineers and quality controllers. Among the latter in particular, opinion of management communications, whether on task-centred subjects or in respect of their market position, was bad, and even very bad. Access to management and supervisors was poor, and their opinion of the human relations skills of management was low. Trade-union membership among these groups was in fact more significantly related to these latter facets of industrial relations than to market facets. Among draughtsmen and laboratory technicians there was greater satisfaction with the work-centred aspects of their job: their dissatisfaction focused upon market factors and with group or occupational status within these wider contexts.

In the view of many theorists, from Marx onwards, the process of 'homogenization' in the market and in the work-place is likely to be a major factor in the realization of a collective sense of identity by those occupying a similarly subordinate position in the formal hierarchy. Therefore one might deduce that the reverse might also be true, and that any influences which act to reverse the effect of major variables enumerated above might also affect the propensity for unionization. This is rarely stated explicitly, but there is sometimes an implicit assumption among management that such conscious modifications to the work environment as those which bring about 'job-enrichment', or informal and 'organistic' work-groups, will increase work satisfaction and thereby reduce the perceived necessity for external representation.[1]

One of the ways to test whether behaviour is patterned in a bureaucratic manner is to test whether it is perceived to be so structured by the actors themselves. It does not necessarily follow that a physical division of labour will lead on to the formal and affectively neutral style that is commonly associated with centralized authority and extended lines of communication. Nor does it follow that small work-groups must be managed or supervised in an informal and personally specific manner: a bureaucratically coercive style can be adopted by the supervisor of a few people as of many. Therefore it was felt useful to ask the sample if they felt that 'in your group or department, management treated you as an individual or as part of a grade or category'.

Table 8.16 Technician views on treatment by management

Perceived treatment by management	Laboratory staff		Draughtsmen		Quality controllers		Planning and production engineers	
	Trade unionist	Non-unionist	Trade unionist	Non-unionist	Trade unionist	Non-unionist	Trade unionist	Non-unionist
Very much as an individual	7	16	8	8	17	21	10	19
Usually as an individual	29	30	26	29	33	23	45	36
Varies	25	16	21	21	15	20	20	13
Usually as a category	20	20	28	24	20	20	15	23
Very much as a category	18	12	14	13	13	14	10	6
Total	100	100	100	100	100	100	100	100

NOTE: Includes 'don't knows'.   SOURCE: Questionnaire Survey

As can be seen, management's treatment was somewhat more likely to be seen to be individual by line technicians if they were in unions, while laboratory technicians and draughtsmen were rather more likely than the other groups to feel the effect of 'bureaucratization', whether in unions or not.

This result concides with the spread of opinion on the state of departmental morale: unionists among quality controllers and

[1] J. Child, *British Management Thought: A Critical Analysis* (Allen and Unwin, London 1969). See also G. S. Bain, op. cit. and Lloyd Reynolds in *Labor Economics and Labor Relations* (Prentice-Hall, Englewood Cliffs, N.J., 1956), pp. 169–77, for an exposition of the managerial strategy of 'peaceful competition'.

planning and production engineers were significantly more likely to feel morale to be high or even very high than non-unionists. This contrasts with the 45 per cent of unionists in the drawing office who thought morale to be low, as against only 22 per cent of the non-unionists who held a similar opinion. It is unwise to make too sweeping an interpretation of the differences between these groups. Yet from interview data it seems that unionization can have an important positive effect on the way technicians view upwards communications in respect of grievance procedure, particularly among line technicians. In addition, across all groups unionists were significantly more likely to regard industrial relations between technical staff and management in their firm as comparing well or very well with that in other companies.

Unions could be seen to enhance the work climate by their members, and, as a means of expressing grievances, they were regarded as valuable supplements to direct personal relationships with management. Line technicians gained a special sense of personal status through the introduction of a union-based grievance procedure. The more fundamental discontents of laboratory and drawing-office technicians were more likely to be seen in terms of promotion and salary levels. While being regarded as an important influence in determining these matters, it was clear that unions did little to alleviate the level of discontent in these areas of concern for technicians. In the nature of things, they generally appeared to raise the level of *expressed* discontent about salary.

The effect of the work-group

It is perhaps significant that there was an apparently widespread lack of knowledge of industrial relations in other companies and an unwillingness to make comparisons. Nearly a third of quality controllers and of laboratory assistants signified that they 'did not know' how the state of industrial relations in their own establishments compared with elsewhere. The full implications of this lack of knowledge of the external situation can only be appreciated if the process of unionization is seen as taking place within and around the local work and market situation as perceived by the potential member. In this process the impact of intra-group interaction on the views of potential members is extremely important, especially where the group is defined in space and function as it is in the drawing office or laboratory.

Between 56 per cent (quality controllers) and 69 per cent (planning and production engineers) of all technicians work in groups of ten or fewer (about 5 per cent working alone), and this group size appeared unaffected by the production technology of the firm. Despite the widespread belief to the contrary, small groups seemed to have no significant negative effect on attitudes to trade unions: indeed, both draughtsmen and laboratory assistants were significantly more likely to belong to unions if they worked in very *small* groups.

In order to understand this we must revert to the fact that 70 per cent–75 per cent of our sample were aware of a union for which they were eligible as existing within the plant when they joined. When the informant knew of a union in existence in the laboratory the chances of his becoming a member were roughly 2 in 3 (66 per cent). When the informant was not aware of a union only 28 per cent reported themselves as being currently union members. In the drawing office 59 per cent of those who were approached by a union representative when they joined their section in fact joined that union, as against only 4 per cent of current card-carrying unionists who had not been aware of a union within the firm when they had first entered it. Among quality controllers, the proportion of 'approached' members fell to 48 per cent, as against 17 per cent who brought their union card into a non-union shop. The proportion of the former type of member among planning and production engineers fell to 40 per cent, as against 7 per cent 'pioneers'. The explanation for the small number of 'card-carriers' between jobs in these latter two occupational categories has already been suggested in terms of the career promotion implied by the move.

Our sample were asked 'If you have ever joined a trade union for technical staff, how many people would you say were already members in your section when you joined?' It appeared arithmetically probable that the smaller the work-group to which most techicians were attached, the more likely that existing union members would make up a higher proportion of the group. In fact it emerged that those who joined unions on entering a firm were most likely to be those who had found unionists to make up a majority in a small work-group of ten or fewer. This was the experience of 52 per cent of laboratory technicians,

59 per cent of draughtsmen and 20 per cent of planning and production engineers who entered such small work-groups. It was much less likely to be true of quality controllers; 55 per cent of these worked in small groups of 2–10, but only 11 per cent of this number found the majority in a union when they joined. However, quality controllers were rather more likely to work in larger groups of up to 20 than other technicians, and the significance of the prior existence of a union in the place of work was high even among the latter group: 48 per cent of unionists in this group found at least half of their section in a union upon joining their firm. This compares with 79 per cent of draughtsmen unionists, 70 per cent of laboratory technicians in unions, and a low 36 per cent of planning and production engineer unionists.

It seems, therefore, that for most draughtsmen and laboratory assistants, joining a trade union may be a 'social' act, and has to be distinguished from the kind of rational choice which seems implicit in the *a priori* categorization of attitudes by the union or professional membership. Whatever the ideology of the individual draughtsman or laboratory assistant, the short-term psychic costs involved in a rebuttal of what is normatively acceptable in his work-group seems likely to outweigh the long-term ideological commitment expressed in his personal career ambitions.

For the minority of planning and production engineers, as for the much larger proportion of trade unionists among quality controllers, union membership appears to have demanded a much more conscious decision, taken either before recruitment to their present firm, and therefore representing a longer-term commitment of a 'moral' kind, or to have involved participation in some form of 'protest' movement, often culminating in a strike.

The exposure of planning and production engineers and quality controllers, in both social and market terms, to the response of management has already been emphasized in the relationship observed between their union membership and the attitude of management to such allegiances. The latent discontent brought out in Tables 5.7 and 8.13 sometimes emerges in an explosive manner among quality controllers. The work and market situation of the planning and production engineer makes it even more costly for these highly status-conscious technicians to take

collective remedies for their clearly perceived grievances, nor
were they often perceived in terms conducive to collective
remedies. The problems defined as 'issues' between the line technicians and management often arose out of role-conflict situations, from personal rebuttals in a face-to-face situation, all of
which were heightened by their desire to be 'accepted' as part
of the management team.

Strikes

Strike action over a wage claim was the only issue to derive
support of any significance—and other forms of collective action
were rarely mentioned. From the figures reproduced in Table
8.17 one can gain some notion of the high degree of militancy
among draughtsmen as compared with other groups. More
interesting is the fact that despite the high density of union
membership among laboratory assistants, their past willingness
to take militant action is less marked than that of either draughtsmen or quality controllers. This may be partly a function of
time in the job, since many laboratory technicians had been at
work for two years or less.

The propensity to strike over wage issues in the past did not
prevent a current desire for 'professionalization'. Of draughtsmen who regarded membership of professional bodies as
relevant, 35 per cent said they had been willing to strike; of
laboratory assistants 45 per cent; planning and production
engineers 16 per cent and quality controllers 11 per cent. Again
we find evidence of 'professionalization' providing an equally
acceptable market strategy to that of unions for the more 'cosmopolitan' laboratory assistants and draughtsmen, but one which
does not preclude the most militant form of collective action.
'Professionalized' laboratory assistants are, for example, more
likely to have been willing to consider strike action over a wage
claim than trade unionists taken by themselves (these are not
mutually exclusive categories). On the other hand, more
company-oriented quality controllers and planning and production engineers tend to dichotomize between what is appropriate
to 'professional' behaviour and what is clearly union action,
again suggesting that 'professionalization' may not be a uniform
phenomenon at all, but one deriving meaning from the actor's
situation.

Table 8.17 Since you have been a technical staff member, have issues arisen over which you were willing to go on strike?

Reasons for striking	Yes or No	Draughtsmen		Laboratory staff		Quality controllers		Planning and production engineers	
		Trade unionist	Non-unionist	Trade unionist	Non-unionist	Trade unionist	Non-unionist	Trade unionist	Non-unionist
Over a wage claim	Yes	62	10	32	15	39	15	30	15
	No	24	69	59	68	50	66	55	72
Over a holiday claim	Yes	8	5	4	3	6	0	5	4
	No	58	70	80	74	72	75	75	79
To get your union recognized	Yes	13	3	3	2	7	1	5	2
	No	54	72	80	78	72	71	65	79
Over overtime payments	Yes	10	2	6	2	4	5	5	0
	No	59	72	78	74	74	71	65	83
Over disciplining of colleagues	Yes	1	1	4	6	7	3	10	0
	No	63	74	77	75	72	73	60	83
Total for each reason[1]		100	100	100	100	100	100	100	100

NOTE: (1) Includes 'Don't knows'.

SOURCE: Questionnaire Survey.

K

The interview material gives more precise guidance to the nature of strikes or other forms of collective action such as 'go-slows' or 'working-to-rule' among technicians. From this it is apparent that such action is most likely to accompany the introduction of a technician's union into a firm, rather than to be a continuing feature of the subsequent collective bargaining relationship. This discovery, which seemed consistent across many different organizational structures, technologists and market situations, does *not* appear in the above table. The reason for this is because the only strike which could be isolated as being purely in support of the recognition of a technician's union by management (what one might describe as a 'principled' strike since it usually involves a direct confrontation with manage-ment[1]) was in the steel plant which provided our pilot study. (These figures are not included in the tables). For most informants such a 'principled' confrontation with management did not come within their frame of reference. Instead, strikes or other actions accompanying the introduction of a union into the situation were almost always perceived in terms of specific substantive issues—usually a wage increase or other improvement in condi-tions given to shop-floor workers or other groups of staff workers, but not to the protesters.

Of recent date such 'principled strikes' were largely confined to quality controllers, though sometimes involving groups who might well have been described as laboratory staff, since they were involved in chemical analysis. The most recent one was in fact in support of the previously mentioned attempt by a group of quality controllers to gain recognition for the Association of Supervisory Staffs, Executives and Technicians (ASSET) in place of the 'industrial' union recognized for white-collar staff

---

[1] The term 'principled strike' is used here to describe collective action which brings no direct benefit to participants in terms of immediate increases in remuneration or improvement in material conditions of work. The most impor-tant issue in this form of dispute is usually union recognition, but it may be used for strategic gains in improved procedural arrangements for collective bargain-ing such as the closed shop, particularly pre-entry closed shops. The significance in the use of 'principled' as a description is to be found in the degree of com-mitment required of a member, and the longer-term perspective of the situation which enables the establishment of a union to be seen as a short term end in itself. Only a person with high normative involvement in the affairs of the union or of the work-group is prepared to engage in such 'pioneering work', so that more substantive issues must be exploited by activists in order to engage others in collective action.

in the steel industry. They have to be contrasted with the 'calcu-lative' goals of draughtsmen in strategically planned strikes, often centrally directed from the headquarters of the Draughtsmen's and Allied Technicians' Association or by the Association's officers at company headquarters. This is not to say that the involvement of all the strikers was always of a similarly calcu-lative kind. Indeed, the 'moral' commitment of many of its local representatives, especially union 'activists', had enabled union leaders to evolve such a bargaining strategy (see Chapter 3).

The translation of a diffuse strain in work relationships into a group consciousness was rarely accomplished in terms of conflict over long-term issues, such as recognition or the procedural arrangements for consultation, but rather was seen in immediate market issues. The knowledge of the procedural elements of recognition among members who were interviewed a long period after the initial establishment of those procedures appeared to be slight, save among a small nucleus of activists. Even among these often well-educated and relatively sophisticated unionists, interest in the union generally implied little desire for a wider commitment to the union *per se* and to its wider social interests.

The process of unionization

Four situations in the growth of a union can be distinguished. In the first the union is either not initially present in the situation or is present only in the person of an isolated long-term 'card-carrier'. Given that the work situation is conducive to collective interaction, the existence of strain within the situa-tion, which presents a challenge to the normative concept of the authority-relationship within the workshop or firm, may lead to some sporadic outbreak of collective or socially expressed dis-content. This very often takes the form of a work-to-rule, deliber-ate obstructionism, or perhaps a 'round robin' or deputation to the supervisor or departmental head.

There are many constraints to such action, not least the feel-ing of loyalty to the supervisor.[1] Most often in the survey section heads seemed to be more oriented to the work-group for whom they were responsible than to administrative management. This

---

[1] Throughout this text the term 'supervisor' is used to include technologists and scientists in a first line overseeing capacity, as well as lesser-qualified chief inspectors and others.

was not, however, always so straightforward a relationship as some texts tend to suggest. In one small-batch research and development establishment the normative reference group of both supervisors and local management was that of a scientist professional. Their more 'cosmopolitan' outlook had caused them to reject the claims to recognition of the lesser-qualified and more 'locally' oriented technicians. The normally amicably 'organic' arrangement of work was therefore periodically disturbed by eruptions over lines of status conflict.[1]

In general the constraints were greatest upon those whose market position was most marginal. Hence there was a high level of latent discontent among quality controllers and planning and production engineers. The more highly qualified laboratory technicians and the ex-apprentice draughtsmen did not feel these constraints to the same extent, and were much more likely to use manual workers as an example of the efficacy of collective action. For these people their status was not really in doubt; but their 'just' rewards were.

The desire for status among quality controllers and planning and production engineers may go beyond relative material rewards to a desire for personal and social recognition by management. As a consequence, unionization among these categories of locally oriented technicians is rather less probable than among the former. If and when it takes place it is usually accompanied by collective action of an explosive kind such as a walk-out or a go-slow. Even these may occur at frequent intervals without evolving into a 'continuous association of employees'[2] of a formally institutionalized kind. It requires the existence of a nucleus of 'card-carriers' with some ideological or normative commitment to the union to bring about internally self-sustaining unionization. More often it was found that local leaders in such eruptions had defined the source of the strain in terms of a need to introduce a countervailing form of representative body into the situation, and had therefore sought the aid of a full-time trade-union organizer from outside the firm. In these circumstances the choice of union was often a matter of chance —governed by acquaintance with someone involved in the fracas, a union already organizing manual workers, the most

[1] See J. V. Eason and R. Loveridge, op. cit.
[2] See Sidney and Beatrice Webb, *The History of Trade Unionism* (Longmans, London 1920).

immediately available union officer. An important element in this choice was the refusal of many manual unions in the firms within the survey to take such recalcitrant technicians into membership. These reluctant organizations included ex-craft unions such as the A.E.F. as well as general unions such as the T.G.W.U. The former refused on the apparently self-defeating grounds that they had no negotiating rights within existing procedures for technician or supervisory membership; the latter because such a step would involve too big a break with existing negotiating interests.[1]

In the case of more highly concentrated laboratory technicians and draughtsmen, unionization was much more likely to grow outwards from the introduction of a 'card-carrying pioneer' cell of membership. The process could be precipitated by some action of management which brought home to technicians their exposed market situation, but it was rarely of an explosive kind. For example, when in the course of a 'rationalization' a company management paid off six senior engineers in a small research and development establishment, union membership in the plant immediately increased from about 10 per cent to 50 per cent. Without resorting to strike action the union (A.Sc.W.) gained management recognition shortly afterwards and continued to negotiate on behalf of technicians in the laboratories despite a relapse to only 20 per cent membership in that plant.

As a result of the premium which these technicians believed should be set on their qualifications, their concern with wider market issues was focused upon retaining this value. In this type of unionization, choice of union style was important. Status considerations affected the character of the organization that was chosen in the first instance: it was a choice taken in a much more considered and 'rational' manner than that taken in an explosive situation. For that reason, even though the union fails to retain its membership, it remains more truly 'representa-

---

[1] This could have been a very good reason in the circumstances in which such a general union was attempting to represent a large horizontal band of lesser skilled workers with extremely limited resources. Since the time of our survey both the A.E.F. and the T.G.W.U. have completely reversed the attitudes displayed in this earlier refusal to recruit technicians. However, as is brought out in Chapter 3, this policy reversal has occurred at national level, and its effects on the present recruiting efforts of their local representatives have still to be revealed.

tive' than the more general or manual workers' union, which may become the vehicle for protest in more anomic conditions. However divorced members or former members may feel from the continued ideological expression of protests emanating from the media of their national headquarters, they tend to regard their representatives at local level as part of their work community. Their full-time officers must be in touch with their problems and able to use the technical language of the craft or profession.

On the other hand, it is much more difficult for the union which has been chosen simply on grounds of ease of access and which has no specialized interest in the particular field of employment to remain as a truly representative body. The third form of unionization among technicians is, therefore, that which stems from a break-away from a long-established 'industrial' or 'general' manual union.

In the case of the 'industrial' union, the shared environmental background of manual workers and technicians (often related by family ties) had contributed to a company-based orientation to work. However the inability of the manually based union to satisfy technicians' interests had led to their growing frustration, and eventually to an explosive walk-out over the 'ascendancy' which the manual workers had in union affairs and in the work situation. Similarly, in the two oil refineries included within the study, the inability of a general union to retain either the earnings differentials or differential career opportunities for low-qualified and company-oriented laboratory assistants during the negotiation of a productivity deal had caused a protest walk-out. In both instances the union which benefited was the one which sold itself on its ability to offer specialized services across a range of technical occupations, ASSET (now the A.S.T.M.S.).

The fourth type of technicians' union is the one which evolves from a 'professional' body interested in maintaining work standards, to one in which the 'just rate' for the skills on offer becomes a matter for bargaining. Chapter 3 outlined the historical development of such 'protective associations' among engineers. This process was not encountered in our survey, although it was proceeding throughout industry at large at the time that this study was undertaken. Instead, what was discovered was that professional bodies were far more desirable means to personal status and the concomitant market security for the more locally oriented, company-trained technician, whose skills, either because of training or age, were specific to the production process of a firm or who was in some other way unable to move between firms. These people were more likely to find the opinion of management an important constraint on joining a union. In these situations management were clearly in a far stronger position to shape the structure and type of employee representation, and to encourage the establishment of staff associations with a more 'professional' emphasis.

On the other hand the obvious involvement of employers in the formation of staff associations contains a large element of risk for the employer. Not only will it leave him exposed to a possible legal action under the Industrial Relations Act 1971, but his actions are open to comparisons made by the whole of his staff, acting as a group rather than as individuals. In other words he has helped to create a network of communications and a means to intra-group comparison where none existed before. Given this new means to group action it takes little time for a group consciousness to emerge. Should this be the case then any failure on the part of the employer to maintain group status and earnings *vis-à-vis* that of another group, say shop-floor workers, or to treat the staff association as anything less than a fully representative body, can result in an explosive withdrawal from the employer-designed machinery of consultation. A possible fifth method of unionization may therefore be created by the employer himself. This was discovered to be the case in one large oil company contained within the study, but it has subsequently emerged as a major means of unionization among employees who are heavily dependent upon company-based career structures and local labour markets, for instance insurance clerks. These employees find the need for an autonomous and nationally based union a more real insurance against the basic market insecurity felt by staff in multi-plant (often multi-national) companies.

Yet there remains a major area of work-centred needs in which the technician's desires for self-achievement and for group status are being increasingly blocked. At the present time the ability of the locally trained and formally badly qualified technicians to join professional associations is constrained by the entrance requirements of the newly formed Chartered Institute of Engineers. This appears to be in the process of rectification, but the history of other protective associations of a professional variety suggests that the establishment of a universal qualification for technicians may lead to a similarly widespread demand for its recognition by management. The course of the subsequent evolution of the new technicians' institutions, and the tactics they adopt, will clearly be determined as much by management as by the technicians themselves.

9 Management Policy

In previous chapters the attitudes and behaviour of technicians have been analysed with reference to factors within the immediate work-place situation and to those external to it. The role of management has so far not been examined separately as a key variable both influencing and being influenced by the behaviour of technicians.

The extent to which management has influenced the behaviour of technicians through its policies is clearly a factor of critical importance in this study. At one extreme, it was suggested by some technicians interviewed that management policy was the factor mainly responsible for the growth of discontent and for their swing towards militant trade unionism. At the other extreme there was strong evidence that many technicians still identified their interests more closely with management than with manual workers, and were strongly opposed to adopting trade-union methods as a means of overcoming their frustrations and dissatisfactions.

There is a problem in discussing the role of management in relation to technicians which arises from the fact that management is a variable element. 'Management' for technicians employed in a small research and development plant, placed geographically hundreds of miles from administrative headquarters, is something quite different from 'management' at another plant where technicians are employed in large numbers in close proximity to thousands of manual workers and to the headquarters of the company. The relation between technicians and the managers who are in charge of the departments in which technicians are employed, who are often technologists, is normally one of mutual respect. Often, however, there is the feeling among technicians that managers with graduate qualifications show a greater concern for those beneath them who have similar qualifications, and they resent what they see as managerial favouritism of graduates. The separation between this level of technical management, with which technicians often closely identify themselves, and the administrative heads who take the crucial marketing and financial decisions, is a factor from which stems much of the insecurity and frustration apparent in the technician's response to this inquiry.

We have shown in previous chapters that the attitude of technicians to management depends greatly upon a variety of factors, including the size of the group, its degree of isolation, the nature of the technological circumstances, the level of salaries and conditions of employment, the opportunities for promotion, and the extent to which alternative jobs are easily available.

The attitude of senior management in general to the technicians' problem was one of uncertainty, ambivalence and often complacency. There were some managers who were prepared to concede that the behaviour of technicians was changing and that they now had to face the prospect that technicians would act like manual wage-earners by adopting a collective bargaining approach to finding a solution to their problems. Most of the senior managers interviewed were, however, reluctant to believe that technicians would see their interests as best served by a close alliance with manual workers. The view of the majority of management was that there was little that could be done about the general technological and economic factors that were creating some unrest among technicians, but that in spite of admitted problems management could still count on technicians recognizing that they shared a closer interest with management than with manual workers employed on the shop floor.

The analysis of the responses of technicians in the plants surveyed gave fairly strong support to these general views of management. In Chapter 5 it was shown that between 50 and 70 per cent of the four main groups of technicians had given an unambiguous answer to the question asking whether they felt themselves to have a closer identity with management than with the workers on the shop floor. It does not follow, however, that technicians are, therefore, likely to refrain from joining trade unions or from adopting a militant collective bargaining policy. A feeling of loyalty to the firm, an even stronger dissociation from manual workers, and a powerful feeling of group solidarity as technicians, expressed in terms of vigorous demands for improvements in terms and conditions of employment, are compatible and complementary ingredients of a rational behaviour pattern.

The general attitude of senior management to technicians was determined by a variety of factors within and without their immediate industrial environment. Apart from the personal predilections of those interviewed, which were a product of their own upbringing, personal experience, and the past industrial environments they had worked in, managers were most clearly influenced by the policy of the company for which they worked. The policy was often influenced by the hunches and beliefs held by members of the Board, and in particular by chairmen and managing directors, who, their subordinates often believed, were not well informed or were impervious to the information which they received from line and personnel managers.

Information flowing to senior management was not always consistent or adequate for an effective understanding of the situation for which they were making important policy decisions. Line management in close touch with different groups of technicians almost inevitably perceives the situation differently from a manager in a personnel function. Reconciling these different views of the situation is always likely to be difficult; it is especially so if senior management is in a position of fundamental ignorance.

To let sleeping dogs lie was a dominant element in the philosophy of perhaps most production managers met within this study. Most of them were preoccupied with problems connected with their technological and production responsibilities, but many were also prepared to recognize that there were latent problems and even agreed that in future there could be considerable difficulties as the position of technicians changed. 'But why bring trouble now?' they asked. It was better to cross bridges when one came to the ditch. To start probing into problems before they had manifested themselves in production slowdowns and stoppages was to invite them.

The problem was made clear in the course of discussion with the management at two companies, both major employers of technicians, which were not, however, included in the enterprises covered by the survey. In one case the senior personnel managers were convinced that there were serious problems arising from the employment of large numbers of technicians, but line management, preoccupied by production problems and seeing technicians only in those terms, refused to recognize that they had much significance. Line managers were confident that they would not have any strikes, and that if they did, these would not have any immediate effect on production. Moreover, many line managers were of the opinion that the surge in the militancy of technicians and their joining of unions had been induced by the role of the personnel department and alleged interest of personnel officers in collective bargaining.

Line managers not infrequently held the view that there were no real industrial relations problems among technicians that could not be solved by a 'straight talk'. Holding this view, production managers were often extremely hostile to any investigation of the factors influencing the behaviour of technicians. This view was also frequently conveyed to senior management, which was often prepared to accept its validity. It was believed by many production managers that this survey of technicians would 'put ideas into their heads', that it would encourage them to think about joining trade unions and making militant demands.

The attitudes of the different levels of management to technicians, while having much in common over all the plants surveyed, were influenced in different degrees by the following factors:

(a) Size of plant and company
(b) Type of production technology.

The effect of size on communications

In large concerns with an advanced technology in aircraft manufacture, electronics and automobile manufacture, the number of technicians employed tended to be in the hundreds. They were also employed in large numbers in the steel and engineering industries. Where technicians were working in units of a considerable size in the firms covered in this survey, there was an inevitable tendency on the part of management to treat the technicians as a category of employees with standardized terms and conditions of employment. Nevertheless, there were considerable differences between the different groups of technical staffs in their relation to their section managers. The significance of these differences has been closely examined in earlier chapters. What is of importance here is that senior management was often little aware of them.

In some firms employing technicians on a large scale there was a close relationship between the technicians and their departmental or section head, against company policy. This common interest was sometimes based on an understanding derived from a shared background—the section head having formerly been a technician. More generally it was based on a shared professional interest in the work being carried out, and a departmental unity in the face of pressures from finance- or production-dominated line management.

In the small departments and small concerns the intensity of professional feeling was usually very great. It was often such that the views of the section head carried great weight and influence with the technicians under his supervision. However, where for personality reasons disharmony reigned, technicians were in a difficult situation. If, for example, they joined a union in face of the hostility of the section head they might find themselves in an uncomfortable isolation and exposed to pressures the union could do little or nothing to relieve.

The effect of technology on managerial communications

Unfortunately it was not possible to examine systematically the attitudes and behaviour of management towards technicians in relation to different types of technology. The response of technicians to a question on their ease of access to management suggested that this was good in most of the plants surveyed, and that technological differences had relatively little effect on this situation.

When the answers to questions on communications with management on a range of issues of concern to technicians were analysed by Woodward's technological typology, namely unit/small-batch, large-batch and process, they showed only minor differences. Technological factors may well, however, have had a major effect on the organizational structure of the enterprises employing the technicians, and this in turn could well have been a factor of great significance.

The differential effect of technology might well have been cancelled out by the development of management policies designed to counter the problems of communication created by technology. It was evident that the exceptional attention paid to personnel management techniques, and to the fostering of human relations skills by the companies which had large-batch and mass-production systems, had been stimulated by a recognition of the need to offset the adverse effects of these types of technology. Although a number of the plants with a large-batch and mass-production type of technology belonged to companies which had special interests in developing personnel management techniques, some of the plants in the unit or small-batch technology category, where the level of criticism was higher, were also distinguished by their concern for good industrial relations. Unfortunately it is not possible to say with any degree of certainty how far technology made management's task particularly difficult. It will also be recalled that it was suggested earlier (in Chapter 5) that the impact of technology on the different groups of technicians may have varied in its effect on their ability to form easy social relationships with their colleagues, supervisors and managers.

Particularly in drawing offices and also in laboratories, this ability was fostered by the extent to which the nature of the work required working groups to be composed of individuals from different levels of the status structure of these departments; it was also fostered by a physical remoteness from the production areas which encouraged something of a community spirit. Even where the distance from production areas was small, an almost academic atmosphere was often established which engendered not only a common bond but also a positive feeling of separateness from the plant. Work also tended to be a little more leisurely than that in the other two main groups, either because of its creative nature, as in the drawing office, or as in the laboratories, because it was not normally too closely tied up with production schedules or targets. In this situation managers were able to take more time in maintaining close contact with their subordinates.

The quality services technicians in particular, and also the planning and production engineering technicians in some cases, often did not enjoy the advantages of the draughtsmen's situation. Of the smaller occupational groupings, instrument technicians in particular were in the same situation as the quality service and planning and production engineers.

Furthermore, it was also mentioned above that, in the organization structures of the plants, quality services, planning and production engineers and instrument maintenance staff all fell under the production or operating wing, whereas the drawing office and laboratory fell under an engineering services or technical services wing. There were clear-cut examples of this division in the steel plants, and of the possible results which could arise out of it. Those technicians who came under a production or operating division felt that a common managerial opinion of them at that level was that they were useless to the central task. This was probably connected with the fact that in comment on the questionnaires, quality services technicians particularly alleged that communications with their section manager were adequate, but that communications with departmental managers or divisional managers were made ineffective because of a lack of interest on their part. The time and interests of these managers were largely taken up by the problem of production or plant operations.

Therefore, it may be that such factors were the major obstacle to making communications satisfactory for technicians, especially for those in quality services, industrial engineering and instrumentation. However, in the plants surveyed managers had sought to tackle the problem of communications for technicians, and any such built-in difficulties, with varying degrees of effectiveness; it is, therefore, necessary to ask what faults or qualities of these different managerial policies and attitudes were the most significant.

Methods of communication

There were normally three main methods by which management attempted to communicate with employees: by written means, for example a plant news-sheet; by union representatives and committee meetings with staff representatives; and by direct contact between employee and manager or supervisor. However, in some of the plants in the study it was found that technicians were not kept in regular touch with higher management through any of these three methods, for a variety of reasons. The number of employees at the plant might be too small for management to feel a news-sheet was justified; there might be no staff union representing technicians in the plant, and, even if there were, staff representatives might not be met in committee because this was felt to be unnecessary; and since technicians were employed throughout the plant it was difficult for them to be kept informed of any important news, plans or developments, by regular contact with the appropriate manager.

Management could often put forward plausible, apparently considered, reasons why nothing was being done to improve communication arrangements with technicians. However, management thinking often seemed somewhat muddled and indecisive on this aspect of policy, and there was often an alarming lack of awareness of the extent to which technicians were critical of the lack of communications, and managerial attitudes towards improvements in this respect. A principal belief of management was that a technician, as a staff member, preferred individual treatment to being dealt with by a bureaucratic procedure, perhaps involving committees with staff representatives, agenda, formal meetings and joint discussions between management and staff. This had been established management thinking in the days when the plants in the study had employed only a handful of staff, and before technical change had created the need for large numbers of technicians. Managers quoted the argument about the importance of individual treatment, but on reflection, confessed that it was no longer possible. For shop-floor workers communications were said to be much easier, for there could be weekly discussions with stewards and regular works committee meetings. However, some managers seemed to be making life unnecessarily difficult for themselves by a reluctance to consider anything vaguely similar for staff. In some plants, it was neither the practice nor the intention of management even to have representatives of staff, including technicians, on joint consultative bodies. One of the reasons for this policy was doubtless the fear that any such regular meetings or the election of staff representatives might lead to technical staff unionism.

In some of the plants where criticism of communications was comparatively mild, this seemed to be due to management policy coping with the problem in a realistic manner. In one case, that of an electronics plant, service technicians were required to work away from base a good deal, and some of them were also expected to work from home; it was often difficult for these technicians to spend very much time in the Service Department or with its managers and planners. Therefore, management maintained a steady flow of written material, especially on technical changes in the product, in order to keep technicians in touch, and regular meetings at head office were arranged for the discussion of any difficulties or grievances.

In another case, that of a vehicles plant, management had organized a plant-wide staff consultative committee, on which technicians were represented, and this committee provided opportunities for regular discussions of problems and for the mutual exchange of views and information.

In one of the process plants, there had been a recent change in the staff consultative committee, from the election of representatives by geographical areas of the plant to election by department. Although this still meant that technicians might be represented not by one of their own kind, but by a technologist, at least this ensured that a manual-worker representative did not represent them. In plants or departments of plants where technicians were densely unionized, managers met the local representatives of their union in the same way that they regularly met stewards from the shop floor, often with a weekly meeting for the mutual exchange of relevant information and views, and the union representatives were asked to pass on any appropriate data. Also, in one steel plant, some departmental managers took the initiative to get together all their department, or, if it was large, representatives of all grades, in a meeting, held perhaps once or twice a month, where there could be free discussion on, for example, the future of the department, changes in product, and problems caused by technical advance.

It is clear that management policy can make a difference to problems of communication, but this is not to suggest that improving communications between management and technicians would be a panacea for all industrial relations problems. What an improvement in communications could do would be to help technicians feel that they were receiving from management the attention that they believe management gives to shop-floor workers.

Redundancy, recruitment and promotion policies

The attitudes of management and technicians towards each other were of particular significance at times when companies were under economic pressure to improve their efficiency either

through rising costs or falling demand. When faced with a need
to reduce the number of technicians employed, management
found it convenient to regard them as a category in a position
no different from manual workers. The technicians, however,
tended to see themselves as a group that should be given a
greater degree of consideration because of their special skills and
abilities and because of their closer identification with management. The ambiguity of the attitude of both management and
technicians was exposed when redundancy loomed.

One company concerned in this survey was uncertain of its
long-term future in its existing location. Although the official
attitude of management had to be one of proclaiming confidence
in the future, the shadow of possible closure could not be entirely
removed. One effect of this situation had been a rise in the turnover of younger technicians. Older employees with a considerable personal social investment in the area were reluctant to
leave or to contemplate moving elsewhere. Trapped in this situation they frequently sought compensation for their frustration
by pressing for improvements in wages and working conditions
that management were reluctant to grant. The consequence was
a good deal of unrest and dissatisfaction.

The dissatisfaction of technicians with their existing jobs and
their readiness to seek employment with another company have
been, discussed at considerable length in Chapter 7. Of the
various factors that were important influences on the behaviour
of technicians, it was clear that uncertainty as to the attitude and
behaviour of senior management was extremely significant. A
substantial proportion of the technicians interviewed believed
that senior management did not understand their particular
problems or show enough concern about their status.

Most of the technicians in the plants covered by this survey
were ambitious for promotion, and they often felt strongly that
promotion opportunities were not adequate. With upper working-class and lower middle-class backgrounds, secondary and
technical-college education, what Olive Banks has called the
'achievement syndrome' was a very real factor for many of them.
They expected to move up the ladder of occupational mobility,
and lack of promotion opportunity, as previously shown, was a
very important grievance. It rankled among those thinking of
leaving their job; and it was also a cause of complaint among
those who were not taking this step.

The reasons behind the promotion discontent varied from plant to plant, but certain points were broadly relevant to a number of them. Firstly, only a small number of companies had bothered to create a graded structure through which aspiring technicians could progress. In the past, when technical staff were not so numerous, there was little need for such a structure, as individuals could be given particular attention. In situations where the personnel of a plant had consisted of managers, a small number of 'staff' and a large number of manual workers, it had been easy for managers to keep an individual watch on the progress of technicians. There was little need to provide the reassurance of a career structure as such, because individual technicians knew that they were being looked after, and since there was a small number of staff in employment it was easy for them to keep in personal contact with their superiors over future prospects. It had been normal for all such staff to have very good chances of attaining managerial status as part of the natural order of things, and, if they did not reach this target, the limited nature of graduate recruitment meant that it was possible to move a good technician to the job of what would today be called a 'technologist', especially when the lower level of technical complexity at that time meant that he could easily adjust to new developments.

It has been shown previously that the vastly increased recruitment of graduate technologists and scientists has been a major factor leading to dissatisfaction among technicians on their career prospects. These graduate recruits often filled the jobs to which technicians had aspired with confidence in former years, and so helped to convince technicians that they would now be stagnating for years to come in the same jobs. Even if a company had a graded structure through which they could ascend, the problem of the graduates frequently remained a great worry to technicians. The fact that in one company, which particularly prided itself on its concentration on the development of human resources, technicians commonly used the term 'graduate barrier' when describing the near impossibility of their moving from the technician ranks, indicated that although movement was possible within a certain structure of grades, there was a level beyond which very few technicians progressed.

Another factor in the restrictions on promotion chances for technicians was the greater degree of job specialization which had affected them. In the development of technical complexity, the growth in size of plants and the increased sophistication of their operations, there had been a need to have technicians specializing in one particular branch of work. In one of the plants, trade-union representatives were pressing for job rotation to give technicians wider experience, but management argued that technical complexity meant it would take too long to train technicians in new jobs and would cost too much. Therefore, the prospects of technicians securing promotion were being significantly reduced by their being unable to get wider experience of work because of greater job specialization.

A factor in this aspect of policy was management's attitude towards the qualifications obtained by technicians. With graduates with good full-time degrees becoming available for employment, and with technical complexity requiring a high standard of qualifications, managers were beginning to think seriously about the suitability for promotion of technicians who had been trained by part-time methods. A number of managers interviewed considered that, even if there were no graduate technologists available, they would be most reluctant to promote technicians to these posts, because of the unsuitability of the way in which they had been trained. As long ago as 1958, Professor Cotgrove[1] wrote that part-time technical training in Great Britain was no longer desirable; it had been a success at the time when the needs of industry were mainly for manual skills. Managers seemed to agree with his point that there was a need for technicians to have full-time training, especially if they were to stand a real chance of being capable of promotion to higher-level posts.

These factors certainly contributed to the frustration of technicians; there were also others which had grown out of a lack of manpower planning. There was a particular problem in plants which had experienced an upsurge in research, since laboratories had been greatly expanded in the last decade, with the result that many school-leavers, all capable of securing the O.N.C., had been taken on. These young people had all expected that they would have good promotion prospects, but they now found they were competing among themselves for the few available promotion opportunities. For with a slowing in expan-

[1] S. G. Cotgrove, *New Scientist*, 20 November 1958, pp. 1308–1310.

sion and with the realization that few people with O.N.C. were suited for technologist jobs, some of the plants had a glut of moderately qualified, young aspiring technicians. The extent to which they felt under-utilized was demonstrated earlier. Current manpower needs were for a combination of good technologists and a low level of laboratory assistants, not for good technicians who might become moderate technologists. These developments created problems for managers and men, which conceivably might have been avoided by better manpower planning.

Companies appeared to have taken on technicians without considering the future growth of the establishment, the potential of the recruits, or their career development, and they seemed to have made little attempt to predict how technical change might alter manpower requirements in a relatively short period of time. One plant had particular difficulties, in that its work was largely concerned with carrying out government defence contracts. Management alleged that they had to take on a large number of qualified technical staff to convince government departments that they were competent to carry out the work concerned, that this meant that they were overmanned on the technical side and that the situation became even worse when the contract approached completion, as this meant that technicians had virtually nothing to do but wait for the next. Among the companies surveyed this was a rather specific case, but the tendency to overman had recurred more frequently.

A common factor which compounded the promotion problem was that of the terms on which school-leavers who were potential technicians were taken on. In particular, the attitude of the companies towards education and training, and towards the level of recruitment, require some comment. On the first point, certain companies with a self-professed view that they had a responsibility towards helping their employees to realize their full potential, or those situated in areas where they had achieved the position of being a local public benefactor, seemed to give the impression that they were promising wholesale education and training. The obvious danger of this policy was that it served to reinforce the aspirations and hopes of their young recruits whose frustration may have been made the harder to bear when they were disappointed. Once young employees had achieved their qualifications they were tempted to feel that they could expect promotion, rather than the stagnation which often resulted from training being given without much consideration of manpower requirements, and their long-term career development.

A suggestion which has been made is that companies may indulge in 'conspicuous consumption'[1] of trained technical staff, in order to add to the prestigious standing of the company. On this point, there was evidence that managers were reluctant to drop the level at which they recruited school-leavers for technicians' jobs, even if they felt privately that they were taking on young people whose aspirations could not be met. It was shown, especially in connection with laboratory staff, that there was a relationship between a high level of academic qualifications and frustration at promotion chances.[2] Management were aware of this situation, but it was still felt that it was better to recruit at too high a level than too low, for it was comforting for them to know that their staff was rather more intelligent than was necessary, and that their department had the high status within the company which often came from the presence of qualified staff. However, in the long term, this situation often seemed to cause worry for both managers and technicians.

There were some managers who had recognized the danger in recruiting over-qualified staff and were deliberately pursuing a policy of hiring techicians who seemed to have a low level of ambition and the bare competence to carry out the tasks required. Those cases were mainly in situations of rapidly advancing technology, which required a small number of highly qualified technologists to make the critical decisions and a large number of low-level technicians to carry out limited routine duties where responsibility was important but technical qualifications of little significance.

In the circumstances, where there was no easy solution to the problem of blocked promotion chances, the introduction of manpower planning and the more realistic assessment of recruitment levels was of vital importance; there were also other policy measures which had been taken by some of the plants to alleviate it. One plant had a large number of draughtsmen from whom there were many complaints about career stagnation, and it was

---

[1] D. Pym, 'Education and the Employment Opportunities for Engineers', *British Journal of Industrial Relations*, March 1969. Pym's hypothesis was not specifically tested in this research, but interview material certainly did not lead us to rule it out.

[2] See Chapter 5.

decided to introduce a graded structure with new job designations and salary levels at each stage. In the new structure there was a basic grade of draughtsman, for the fully trained man who was able to make modifications to existing drawings and to redraw designs from other sources, and this grade was itself divided into two levels, according to merit. Above the basic grade, there were grade 3 draughtsmen, grade 2 design draughtsmen and grade 1 senior design draughtsmen. Furthermore, below the Chief Draughtsman and his Assistant, there were 'senior design engineers', 'design engineers' and 'assistant design engineers', and some of these posts had been created in 1964 at the same time as the new grading structure to provide a graduated ladder by which ambitious draughtsmen could progress. Each year, the draughtsmen were formally reviewed not only as regards salary but also to decide whether they were capable of promotion to the next grade. Although the new structure was rather complex it did serve to give the draughtsmen the stimulation of a career ladder which they could ascend up to the level of designer.

Another policy measure used by management to alleviate career stagnation was to encourage the horizontal movement of technicians into other jobs in the plant. It was only in a small number of plants that this was practised, and it usually took the form either of draughtsmen moving to jobs as engineers in production, or of other technicians moving to become supervisors on the shop floor. However, quite apart from the point that increased job specialization had made such horizontal movements more difficult there were reasons why management was reluctant to expand such developments in the first instance.

Some plant managers were short of draughtsmen, and were unhappy at any of them moving out of drawing offices. In the second, the particular plant concerned, in a move towards demanning, flexibility of labour and removing demarcation differences on the shop floor, had developed a plan for the promotion of shop-floor workers to these supervisory posts, and the mobility of technicians had consequently been curtailed. In interviews, technicians made clear how important this policy had been in cutting down opportunities for them.

One or two managers in several of the plants covered were of the opinion that where no suitable promotion opportunites existed for them, young, aspiring technicians should leave. This advice was being given in a laboratory in a plant where it was felt that there was some overmanning and that there was little chance of promotion for the majority of technicians. In this particular plant, some managers confessed that lack of manpower planning had caused the problem, and there was little alternative for ambitious technicians but to leave.

The assessment of long-term manpower requirements, the relating of these to the future supply of technicians, and promotion and training policies, were obvious needs in most of the plants covered in this survey. Most managers were aware of the problems, but seemed to lack conviction that they could really do something about them.

The improvement in manpower policies would be valuable in the long run; in the immediate future, two possible developments suggest that there might be ways of improving the job satisfaction of technicians. The aspirations of technicians for career fulfilment may be found in the future by utilizing the concepts of 'job enrichment' and 'work structuring'. One of the companies included in the study had been carrying out experiments in 'work structuring' in one of its plants; its feelings that the experiments had been a success made it probable that they would be extended. The experiment was principally aimed at finding out how much more employees could attempt in their job; and secondly, to break production down into small teams, with no foreman, supervisors or rules.

The experiments had been designed to improve the morale and motivation of operators who were bored by unskilled repetitive production work, but a few of the managers within this company's plants thought that the idea might usefully be extended to technicians, for example those bored by inspection work.

The second idea gaining currency is that of 'job enrichment', which may be defined as building into jobs greater scope for personal achievement and its recognition; more challenging and responsible work; and more opportunity for individual advancement and growth. As a management policy, it is based on the theory of motivation of Professor Frederick Herzberg,[1] who is one of the authors of a recent article[2] proclaiming its success. Two

---

[1] F. Herzberg et al., The Motivation to Work (Wiley, New York 1959).
[2] William J. Paul, Keith B. Robertson and F. Herzberg, 'Job Enrichment Pays Off, Harvard Business Review, March/April 1969.

of the five case histories cited by Herzberg were concerned with laboratory technicians and design engineers. The laboratory technicians showed that they were able to write research minutes very nearly as good as those of the scientists they worked under, and the design engineers were placing orders on their own authority for as much as £200,000 worth of equipment. The enrichment of these jobs resulted both in the increased responsibility being carried without any disasters and in increased job satisfaction.

One company with plants that were covered in this survey had developed a specific company philosophy based upon theories developed by Herzberg and other behavioural scientists. The aim of the company was to inculcate within its staff the motivation to accept a high level of responsibility and a readiness to act with speed and efficiency. The fundamental factors which had inspired the company in the adoption of its philosophy of creative participation were related to its technology and its economic position. As an extremely capital-intensive company using advanced systems of process technology it was acutely aware of the need to ensure that its staff understood the critical significance of their role in the success of the enterprise. It was also well aware that in terms of its size, its world-wide importance and its significance to the British economy, it was in an exposed position.

The company had made considerable efforts to win the support of all levels of management for the philosophy and the policies that emerged from it. Not every manager was convinced of its practicality. The real difficulty which the company faced was in persuading its employees of the virtues of the philosophy.

The technicians employed were not impressed by one of the basic features of the companies' philosophy, namely that of a single status for all employees. The aim of the company was to achieve a common standard of pension rights, sickness benefits and other non-salary conditions of employment. The technicians felt that they were being deprived of differential advantages to which they were entitled on the score of their training and responsibility. The advance to parity in conditions by the manual grades was seen as evidence of their own loss of status, and confirmation of a fear that the company no longer regarded technicians as meriting special consideration.

Work structures and job-enrichment policies also gave rise to other difficulties. The trade unions organizing technicians were by no means prepared to see their members taking on extra responsibility and putting more effort into their jobs without improvements in pay. Management might well feel that it ought not to be called upon to pay extra for alleviating low job-satisfaction and improving the security and long-term prospects of its staff. In addition, there can be little doubt that these techniques, whatever there is to be said in their favour, could not provide a solution to the problem of promotion, which technicians were so concerned to see resolved in their favour. What many technicians expected and, in some cases, were encouraged to expect, was that they would no longer be technicians after several years' service, but rather that they would have become technologists. This was only a remote hope for most of them; thus this basic problem remained, to be a challenge to management until some more radical acceptable solution was found.

Salary policy

The salary policy followed by management in the plants investigated was widely regarded by technicians as unsatisfactory. The primary objective of management seems to have been to establish wage levels that would enable the company to recruit and maintain its technician labour-force. However, management have frequently felt it not to be necessary to pay particular attention to the salaries of technicians, but to make general increases following increases negotiated for manual workers. Under pressure from the technicians' unions this situation has been changing.

The factors influencing the structure of technicians' salaries were discussed in Chapter 6. The most important aspect of the pattern of technicians' salaries has been the erosion of traditional differentials in salary levels and conditions of employment which technicians previously enjoyed over shop-floor workers. The dissatisfaction created among technicians by these developments has had a major influence on their attitude and behaviour. Management, while aware of the discontent of technicians and sometimes having a considerable sympathy with their point of view, have generally felt unable to reverse this trend. They have been afraid that any attempt to do so would simply provoke manual workers' trade unions to press for a further increase in pay so as to reduce the widened differentials.

Increases in earnings by manual workers were not resented so much for their own sake, but rather because they brought home to technicians the disparity between the responsibility that they were carrying in their task relative to that of shop-floor workers. Often indeed the line technicians (particularly quality controllers) were, along with supervisors, being expected to carry *increased* responsibilities as a result of changes in work-flow some of which, in two plants at least, came about through attempts to 'enlarge' or 'enrich' the jobs of shop-floor workers. These latter were rewarded with guaranteed levels of earnings, given staff status and generally brought up to the level of technicians in terms of fringe benefits and other status symbols in the agreements which resulted from management-union negotiations. When asked if they resented such an increased status for manual workers in the form of fringe benefits, technicians were in favour of the spread of such rewards to the shop floor. However, once this had occurred it was apparent that technicians had become aware of the fact that the past situation, in which high shop-floor earnings were discounted against job insecurity and fluctuations in bonus payment due to trade recessions, no longer applied. They were now able to compare formally agreed levels of earnings for shop-floor workers with their own, and to go on to make a comparison of tasks and responsibilities and relative positions within the firm.

It was clear from discussions with management that they did not always appreciate the extent and depth of the technicians' resentment against the advance of the relative standards of pay and conditions of employment of manual workers. However, managers in certain of the plants surveyed had made serious attempts to deal with the dissatisfaction of their technicians on salary matters. In one plant it had been concluded that as the earnings of shop-floor workers were felt by technicians to be out of line with their salaries, this problem could best be met by a reconsideration of the wages structure of the entire plant. Accordingly, the plant management had decided to make a systematic attempt to reshape a structure of pay which had developed in a very haphazard way. Wage and salary increases in the past had often come about from the competing activities of manual and technicians' unions and different degrees of

bargaining pressure. The management was seeking to plan the
new structure in such a way as to permit technicians to make
up some of the ground that they had lost. The aim was to
incorporate all grades of employees, up to managerial level, into
an overall wage structure which would seek to establish a pattern of wage and salary levels in which the higher grades of
technicians would be above the highest level of earnings achieved
by manual workers on the shop floor.

The attempt to establish a similar structure of pay for all
employees on the basis of job evaluation in another plant had
resulted in embarrassing difficulties for management following an
analysis of clerical and technical jobs. It had become clear to
management in the carrying out of the analysis that clerical staff
and certain levels of technical staffs were not likely to rank as
highly as had been taken for granted when compared with skilled
manual workers. Management was, therefore, in a dilemma;
since the carrying through of a plant-wide job-evaluation scheme
was likely to promote a serious conflict with the very groups it
was anxious to help. It cannot be assumed that the rationalization of wage and salary structure will automatically be beneficial for all groups of technical staffs. Some groups may gain,
but others may well be in a less favourable situation.

Productivity bargaining may be particularly significant in
the effect that it has on established differentials. Where craftsmen have been persuaded to become more flexible in their use
of restrictive labour practices by the payment of higher wages,
the increase may take them above levels of pay enjoyed by some
technicians. Changes in work practice negotiated through productivity bargaining may also lead to craftsmen extending their
function into areas of job performance that hitherto were regarded
as necessarily carried out by technicians. This type of squeeze
on the work-role of technicians was of much wider importance
than its effect on relative salaries. The concern of technicians at
their relative wage levels was giving rise to a readiness to consider
the application of bonus and merit schemes. The possibility
of linking higher earnings to higher productivity has been
strongly canvassed by the National Board for Prices and Incomes
in several of its reports. In its report on pay and conditions
in the engineering industry[1] the N.B.P.I. suggested that the

[1] N.B.P.I. Report No. 49, *Pay and Conditions of Service of Engineering Workers*, Cmnd 3495, (H.M.S.O. 1967).

problem of pay for staff workers was due to the fact that there was insufficient recognition of skill and responsibility in current wage structures. The Board was of the opinion that general increases in pay would not lead to the establishment of rational and equitable salary structures.

In the N.B.P.I. Report No 68,[1] on two sets of wage agreements made with DATA it was stressed that it was possible for the pay of white-collar workers to be improved by increased productivity. After studying agreements for draughtsmen's pay increases, the Board made four specific recommendations on how the efficiency of drawing offices could be improved.[2] Companies could look at the relationship of the drawing office to their other activities to see whether its work could be more effectively integrated into the whole system of planning, control and communications; draughtsmen could be relieved of routine clerical duties; standards and their control could be reviewed to see whether new techniques would cut the cost of each drawing; and the methods of drawing might be improved through the introduction of simplified or more functional methods which saved time. In Appendix E to the Report, the Board set out in detail possible ways and means of increasing efficiency in the drawing office and of working out productivity agreements.

In Report No 86, on staff workers in the gas industry,[3] the Board made recommendations to produce more efficiency, and these included a proposal for facilitating work-measurement on the jobs of staff. Further, the Department of Employment and Productivity, following study of the problem by its staff and by the N.B.P.I., has urged employers to implement new methods of measuring productivity for white-collar workers.

However, a major problem in the way of adapting these ideas on linking the pay of draughtsmen to productivity is the policy of DATA towards work study. In 1969, the Engineering Employers' Federation published a paper on productivity for staff workers, and it included reference to the problems posed by DATA's refusal to take part in productivity and work-study deals. In an interview on this paper, given to *The Times* on

[1] N.B.P.I. Report No. 68, *Agreements made between certain engineering firms and the Draughtsmen's and Allied Technicians Association,* Cmnd 3632 (H.M.S.O., London 1968).
[2] Ibid., p. 11
[3] N.B.P.I. Report No. 86, *Pay of Staff Workers in the Gas Industry,* Cmnd 3795 (H.M.S.O., London 1968).

14 January 1969, Mr George Doughty, General Secretary of DATA, said that 'it was ridiculous to try to measure a draughtsman's output or productivity by timing him'. In the National Agreement of June 1968 between the E.E.F. and DATA, measures aimed at raising productivity in drawing offices received support, but there was no specific commitment to the use of timing and other measurement techniques. Therefore, in organized drawing offices, the adaptation of measures for improving the efficiency of technicians will be limited by the policy of the union. Consequently, effective methods of improving the efficiency of drawing offices will have to be sought in other directions. In the plants covered by the research, the suggestion that routine clerical duties could be removed from draughtsmen's work was a popular one among managers and men, when they were considering how to improve productivity in justification for pay increases.

To many technicians, management, in its salary policy, was not placing enough emphasis on the possession of individual skills or on the amount of responsibility carried. From the data examined in Chapter 6 it would appear that the reward of individual skills was the point on which technicians felt there needed to be most change in management policy. This was confirmed by the information gathered in interviews in which it was alleged that management did not, as it proclaimed, pay well those who could show a high degree of individual skills. Technicians seemed to feel that management was giving too much importance to seniority as a determinant of salary. The upshot of this situation was that management was being more conservative and less adventurous or modern-minded than technicians generally wished to see. Only the older employees were seen as getting towards the end of their careers the salaries which technicians generally felt they were justified in securing for themselves on the basis of their individual skills and training. The enthusiastic response of technicians to the idea of salary structures based on appraisal of skills and job performance may be seen as a call to management to take more positive steps to reward the individual skills which they claimed were not an important enough determinant of the salaries of technical staff in their companies.

To sum up, the main problems facing management in the field of salary policy were twofold. One was to satisfy technicians that their salary levels were determined on the basis of their genuine worth and contribution to the company, and not because they were less numerous and less militant than manual workers; and the second was to satisfy technicians that the structure of salaries had adequately taken account of such factors as skill, training and the limited opportunities for promotion. In terms of salary for age, technicians wished to see high salaries at younger age levels than were often provided under existing conditions. They would have liked management to match the opportunities which they believed were available to manual workers to secure higher levels of pay through productivity improvements. Although there was strong opposition from some union representatives to any suggestion of directly linking the output of technicians to salary payments, many technicians believed that their skills were not being fully utilized and that with appropriate reorganization they could be paid considerably more without adverse effects on their conditions of employment.

If much of this belief was little more than a pipe-dream, some of it might perhaps have been dispelled if managements had developed closer relations with technical staffs and demonstrated that the problem of technicians' pay and conditions of employment was given as much attention as was given to the pay and conditions of manual workers. However, the basic problem lies deeper than management–technician relations within the firm.

The fears and frustrations of technicians often lay in the wider area of career and market prospects outside any local decision-taking authority. Even when management had shown a considerable understanding of the concerns of technicians and established an effective interaction with them on all matters relating to the technical aspect of their jobs, there still remained a general anxiety about their overall status and future job security.

The structure of ownership in the shape of the multi-plant firm had had a widespread impact upon the attitudes of white-collar employees. The technicians studied in this survey did not feel so subjectively isolated as some psychologists and industrial sociologists have suggested is typical of the alienated worker most likely to join unions. Instead they were aware of their blocked career ladders and their isolation from decision-taking in the geographically distant head-office unit—an isolation which they shared with the senior managers within their own plant. For such managers the problems of communicating with their employees on policy decisions in which they themselves had played little part were extremely difficult to handle. They were forced to take on an entirely different and much more remote role to that adopted in the contrived intimacy of the career-appraisal interview.

It was these factors, as we have shown, that were making for a growing ambiguity in the attitude of technicians towards managers, their own role, and their view of trade unions and collective action.

Recognition of unions

The attitude of management in most of the plants surveyed towards the recognition of unions organizing technicians could be summed up as a grudging acceptance of an unavoidable but rather unpleasant necessity. Most managers would have preferred to settle issues concerning wages and conditions of employment directly with the staff involved without the intervention of a union. Nevertheless, many managers had accepted as inevitable the existence of unions, and though they continued to grumble and complain of the damaging effects of the activities of the unions, they no longer sought actively to discourage union membership or to frustrate the activities of union representatives.

In the past employers have encouraged the formation of staff associations, which were seen as an alternative to unionization. As was indicated in Chapter 3, only one firm with two plants in the group covered in this survey had its own staff associations that included technicians. In supporting the establishment of staff associations the management in this firm had recognized that employees had a need to discuss common problems and to secure collective representation of their group interests. They clearly believed that a staff association was a more appropriate body for technicians than a trade union. In this belief management were without doubt influenced by a conviction that only a minority of technicians actually decided to belong to a trade union. Management was also motivated by the knowledge that a trade union would be more aggressive and difficult to deal with.

Although staff associations were not in existence at the other plants covered in this survey the attitudes of management to the development of trade unions was often not very dissimilar. It was obvious, however, that in some cases a definite change had taken place. The strategy of appealing to the feeling of shared interests which it was known technicians often had with management had in a number of cases been abandoned as unrealistic.

Management in a number of cases had come to the conclusion that they had little to gain by seeking to keep alive a situation which was no longer relevant to contemporary circumstances. It was recognized that in practice technicians had to be treated not very differently from manual workers and clerical grades. Such privileges as staff employees had possessed in the past were gradually being eroded by the advance of manual-worker standards.

While management was well aware that technicians were highly concerned at the whittling away of their superior terms and conditions of employment, they were not prepared to precipitate a head-on conflict with their manual-worker employees merely to satisfy the technicians. There were even some managers who had reached the stage where they saw technicians in the future as they had seen manual workers in the past. This type of manager was prepared to recognize that technicians would join unions, bargain militantly, and perform their tasks efficiently even when treated as no different from any other group of wage-earners that might be employed.

Two main types of recognition were found to be accorded to unions representing technicians. First, wherever DATA was recognized in the plants, it was usually recognized as having full negotiating rights. By 'full negotiating rights' it was normally meant that management dealt with the union both on procedural and substantive matters in negotiating agreements, including the rates of pay for its members, and in settling disputes, or grievances, involving the membership. However, where the A.S.T.M.S. was recognized in the plants, it was usually granted only a limited part of these full negotiating rights. Recognition was extended only to the settlement of disputes or grievances arising out of the terms and conditions of employment and not to the negotiation of comprehensive agreements covering technician grades.

The distinction between these two types of recognition and the importance attached to it by management was evident from the documents of the British Employers' Confederation quoted in 'The evidence of DATA to the Royal Commission on

L

Trade Unions and Employers' Associations'.[1] These documents were concerned with a meeting of the Wages and Conditions Committee of the B.E.C. which considered a summary of observations received from members in reply to a questionnaire on staff workers. A summary of the Committee's views was circulated to members, and in this summary reference was made to the issue of the recognition of staff unions. For example, in paragraph 5 of the summary document, it was recorded that:

> of 23 Member Organisations which replied to the questionnaire, only one was able to state, without reservations, that trade union representation of staff workers is recognised, that this recognition is on a formal basis, and that agreements covering the rates of pay, etc, of staff workers are made at national or company level. A few other Member Organisations indicated that staff unions are recognised to a limited extent (e.g. they are recognised at certain firms or only in respect of certain types of staff workers) but generally such recognition is limited to procedure and does not cover the making of agreements.[2]

Further, after mentioning that there was a growing tendency for staff workers to join trade unions and that some members had said that this was going to require serious consideration of the question of recognition, the summary document mentioned the difficulties which might arise from 'conflicting interpretations of the word "recognition" '. The document went on to say:

> even if 'recognition' were granted to a staff union, this need not include negotiation of wages and conditions of employment, but might be restricted to informal discussions or to the laying down of procedure for dealing with requests and complaints. There is a danger, however, that once a staff union has been recognized, for any purpose at all, as representing the interests of staff workers, it will be encouraged to press for the full rights of negotiation. Those members who have already granted full 'recognition' to staff unions or to the staff section of a process workers' union confirmed that official representation of staff workers could be a source of much difficulty to employers. In the circumstances, the Committee agreed that when discussions were held with staff unions it would be best, if possible, to avoid the use of the word 'recognition'.[3]

[1] British Employers' Confederation, *Wages and Conditions – Exchange of Views and Information. Staff Workers*, June 1964, Royal Commission on Trade Unions and Employers' Associations Minutes of Evidence No. 36, DATA (H.M.S.O., London 1966).
[2] ibid. p. 61–2
[3] ibid. pp. 62–3

From this it is clear that management on the B.E.C. committee at that time saw it as essential to avoid, wherever possible, the extension of full recognition rights to staff unions. Similarly, in the plants covered in this survey it was found that management saw union organization as a useful aid in matters such as the handling of disputes, but that many arguments were deployed against the extension of full recognition rights. The view of management was that limited recognition, although a danger in itself because it could lead to full recognition, was definitely preferable to full recognition.

The arguments employed by managers on why they were opposed to granting full recognition to unions representing technicians fell under two general headings. Firstly, managers felt they were fully capable of looking after the best interests of their technicians without outside interference from unions. This was largely bound up with their self-expressed faith in their competence as managers. Secondly, managers felt that the effect of trade unionism for technicians would be generally bad for both managers and employees.

G. S. Bain has suggested[1] that there were two other major grounds on which management resented the extension of recognition rights. These were, firstly, that a certain union might be said to be not appropriate for the occupation, firm or industry concerned, and secondly, it could be said to be not representative of the employees. No doubt these two arguments may be used by managers as reasons for refusing recognition. However, these two arguments were not commonly met with in this form during the course of our research. There was, however, one instance where each of them was utilized. In one of the steel plants where the A.S.T.M.S. was recruiting members from technicians who had left BISAKTA, and was pressing for recognition, some managers argued that BISAKTA was the union which best understood the problems of the steel industry and its employees. In one plant of a company in the refining industry—not actually the plant covered in this research, but one which was associated with it—the argument of representativeness was currently being used, i.e. that the union had to show it had 51 per cent membership before it could have recognition rights. The argument of appro-

[1] G. S. Bain, *Trade Union Growth and Recognition*, Research Paper No 6, Royal Commission on Trade Unions and Employers' Associations (H.M.S.O., London 1967), pp. 68–97.

priateness was frequently given as a reason why technicians did not intend to join a union. In Chapter 3 it was suggested that some technicians, particularly those in more specialized work, would not join unions because there was none available which was sufficiently appropriate for them. For example, 'appropriateness' was argued by service engineers as a reason for not joining the A.S.T.M.S. or the E.E.T.U./P.T.U., because they felt that their skills were so specialized that they would not get sufficient understanding of their needs from such unions of broad interests. As for the argument of 'representativeness', whether it was being utilized depended on the particular stage of development of the union concerned in individual plants. It is likely that this argument will always be introduced at some stage in this attempt by a union to achieve recognition.

Managers were convinced that they were capable of looking after the interests of technicians without the interference of any outside body; they genuinely believed a union would make close and good relations difficult. They also believed that technicians were being led astray by a minority of malcontents. One of the assumptions on which managers generally based their doubts about the need for effective unionism for technicians was that there was a community of interest and approach shared between managers and men. However, many technicians not only felt that management had not very much interest in them, but were also of the opinion that they were not part of the management team (see Chapter 5).

It has been suggested that the reluctance of management to recognize technicians' unions has been due mainly to an ideological concept of the enterprise as a unitary organization.[1] There can be no doubt that many managers do view the company in terms of an organic institution. However, this was not the predominant attitude of the managers of the firms that were examined in this survey. All the companies covered accepted that their employees had the right to join a union if they so wished. Most of the managers, in both line and personal positions, accepted the need for unions for shop-floor workers, and expected to be put under pressure to concede improvements in pay and terms of employment by them. It was, therefore, not

[1] Alan Fox, *Industrial Sociology and Industrial Relations*, Research Paper No. 3, Royal Commission on Trade Unions and Employers' Associations (H.M.S.O., London 1967).

their general view of the nature of the enterprise which had led them to ignore or misconstrue the interests of technicians, but more particular reasons. Indeed, it would be more accurate to say that their view of industrial organization as a pluralistic system was well established, but it was based on the pluralist system of the past, when the two sides of industry were more simply divided into managers and staff, who were set apart from the mass of shop-floor workers. What managers had been slow to accept was that with the evolution of industry and the advance of technology, technicians and clerical staffs had emerged as separate and distinct interest-groups.

The attitude of management towards technicians' unions changed visibly during the course of this research. As these words are being written it is clear that a growing number of managers are deciding to abandon their previous reluctance to accept technicians' unions on the same terms as they accept manual workers' unions. This change is coming about in part as a result of the Donovan Commission and the establishment of the Commission for Industrial Relations. Outside pressure is not the only factor in this changing situation. Most important are the economic and social factors discussed at length in this study.

There are some major concerns, employing large numbers of technical staffs, who continue to believe profoundly that to recognize unions would be a moral and economic mistake. The managers of these firms hold strongly to the view that the enterprise is an organic unity; that all employees from the top downwards share a common interest in the success of the company and owe it a common loyalty. In return they aim to see that the company provides the best possible terms and conditions of employment and provides the fullest possible opportunities for all its staff to develop their potential abilities and to fulfil these to the maximum extent within the firm.

The managements of these companies do not seek to prevent their employees from belonging to unions, but they tend to believe that to recognize staff unions and to negotiate with them would destroy the concept of organic unity. Although those who hold this view are retreating before the rising tide of staff radicalism, they are often able to maintain the support of a substantial proportion of their staff employees.

Under contemporary circumstances, the trend is towards the organization of technical, clerical, administrative and supervisory grades and their integration into the collective bargaining system. To most technicians this development offers a greater promise of improvement in their status and conditions of employment than any other that management appears to be able to offer in most cases. The evidence from this survey suggests that it is unlikely that the trend will stop short of a much higher degree of unionization and, though reluctantly, militant behaviour.

10 Summary and Conclusions

Technicians do not have a clearly delineated place in the occupational structure. They are generally referred to as if they were a homogeneous occupational group, but this study shows that in fact technicians span quite a heterogeneous range of occupational activities. The technician, under a variety of different job titles, practises varying levels of skills, acquired through a number of educational processes. Although technicians have some features in common with manual workers, there are some clear differences between the two groups. In the production process the technician is concerned with planning, organizing and servicing functions, and may carry an important responsibility for the completion of a particular task.

The education required by a technician generally goes beyond that normally needed by a skilled manual worker, but it stops short of that which is required of a professional engineer, industrial chemist or metallurgist. Although there are clear differences in function and education at the margins, and the margins may be quite broad, the technician overlaps at both ends with the skilled worker and the professional employee. In some of the establishments surveyed in this study employees designated technicians were performing tasks usually associated with those carried out by technologists; they were also found at the other end of the spectrum carrying out jobs that were in essentials little different from those carried out by skilled manual workers.

Technicians have enjoyed 'staff status', and it has been normal for them to identify their interests with those of the concern for which they work. In this respect they have much closer ties to management than to the manual worker grades. However, the relationship has not been comfortable with either management or the manual grades. Technicians have been conscious of the fact that they were marginal men, and have resented their uncertain position in the occupational hierarchy. This ambiguity of position has been reflected in the lack of clear group identity, with the exception of draughtsmen, who have been much more sure of their occupational status and position in the employment structure.

Technicians have educational interests and attainments that place them in a clearly different category from skilled manual workers. Most of them have O level and Ordinary National Certificate qualifications. Their attitude to education is primarily instrumental, seeing their qualifications in terms of career advancement, higher salaries and better conditions of employment. Technicians emerged from this study as an ambitious group of employees, with high expectations of promotion and a readiness to make efforts to improve their technical skills. In these respects in particular the technician felt superior to the manual worker. Individual advancement was seen as desirable and achievable.

Lack of career opportunities and slow advancement due to blocked promotional outlets were widespread complaints. Whereas in the past technicians had been able to move into the technologist grades through job experience, this route was now becoming closed. In the face of changing techniques of production employers were raising their hiring standards and recruiting technologists almost solely from the ranks of university graduates. Technicians view this trend with trepidation, and fear that it bodes ill for their own career development. The 'graduate barrier' has become a factor of major significance to the future of the technician as an intermediate occupational grade.

The situation of the technician appears affected not only by the restriction on promotion, but also by the more efficient manual workers. Programmes of job enrichment and a readiness to let manual workers extend their role has led to a growing overlapping with technicians. Thus, under pressure from the bottom as well as the top of their range of expertise, technicians have seen their future as uncertain and in need of protection from encroachment.

Technicians perceived their earnings to be falling in relation to those of manual workers. The role of Government, in its determination to prevent inflation through an incomes policy, was seen as an important factor in this relative decline. The permitting of wage increases above the norms laid down on the basis of productivity agreements had benefited manual workers, but helped technicians relatively little. Technicians felt strongly that since it was difficult to increase their direct contribution to output the importance of their role was underestimated. Moreover, since they were only weakly organized in trade unions, management had paid little attention to them; technical staff had no 'restrictive practices' to sell and were thus being penalized for their past co-operation with management. Manual workers, on the other hand, were seen as being organized in powerful trade unions, which through productivity bargaining could sell restrictive practices in return for high wages. The relative decline in money wages was being further aggravated by the trend towards the establishment of a common standard of benefits for manual workers and staff employees. The upshot of these developments was the growth of a widespread belief among technicians that workers of a lower status were gaining at their expense.

The position of the technician has been affected by other factors apart from a decline in relative pay. The trend towards larger and larger business units under the force of competition and the search for economic stability has led to increasing bureaucracy and remoteness. Although technicians have been aware of this trend, the personal relationship with management has prevented vigorous protest. Management on the other hand has increasingly come to see technicians as part of the new mass labour force, as a homogeneous group, little different from the manual workers on the pay-roll. Like manual workers, technicians feel they are becoming mere numbers on the pay-roll computer tapes.

In these circumstances it has become increasingly difficult for technicians to retain their previously close identity with management. In the companies covered in this survey bureaucratization had not reached the ultimate stage; the majority of the technicians employed worked in small groups, and their relationship with their immediate superiors was generally good. Even so, complaints of being treated as a category rather than as individuals were frequent. During the prepatory stages of this study two firms were visited where large numbers of technicians were employed, particularly draughtsmen in large drawing offices. There had been a serious deterioration in relations with management and a growing tendency for conflict to occur.

The increasing number of technical staff in employment has meant an increase in the wage costs of technicians and a rising concern with the need to economize this factor. The recent trend towards business mergers has exposed technicians to employment insecurity, and management has sought to reduce its cost burden. In the past technicians enjoying staff status have seen themselves as immune from redundancies. It was the manual worker who bore the main brunt of unemployment. Mergers and nationalization have resulted in management examining closely areas where inefficiency could be eliminated. Technical services supporting production have tended to bear the brunt of such rationalization. The highly publicized G.E.C.–A.E.I. merger in 1967, which resulted in a large number of technical staff being made redundant and the exposure of all to the threat of loss of employment, had a significant impact on the general morale of technicians. The effect of mergers has been to make technicians more aware of the need to join organizations that might protect them from possible adverse consequences. Fear of redundancy was effectively used by the A.S.T.M.S. as the basis of a recruiting campaign in a full-page advertisement in *The Times* headed 'Management has decided it does not like the colour of your eyes'.

As the status of technicians has relatively declined with the raising of the standards of employment of manual workers, they have become increasingly concerned with levels of pay. In short, status has become primarily a matter of relative wages. Technicians have not, however, been able to raise their pay as fast as manual workers. This situation, together with the other uncertainties of their position, has given rise to frustration and discontent among technicians.

Attempts by technicians to overcome their difficulties have followed two main paths: one, an attempt to improve their position by finding alternative employment; the other, to seek salvation by joining a collective organization such as a trade union or a professional association.

Mobility among technicians was found to be small. Many company-trained technicians found difficulty in changing jobs because of the specialized nature of their skills. What voluntary movement there was was largely confined to those with a more general range of skills, who were convinced that mobility was essential to overcome their occupational limitations. Technicians in this category tended to be ex-apprentices, who were strongly motivated to improve their economic position.

There was a significant difference between technicians who wanted to leave their firms and those who did not, in their attitudes to the sufficiency of promotion outlets in the companies they worked for. Those who wanted to leave their employment were much stronger in their criticism of the availability of promotion opportunities than those who were satisfied with their existing jobs. Yet both potential leavers and non-leavers were equally vehement in their condemnation of the inadequacy of current wage differentials between technicians and manual workers. Although technical change was altering the level of skills required of technicians, there was little relationship between the desire to secure a better job and dissatisfaction with the utilization of the skills of technicians by the company concerned. Labour turnover of technicians was, of course, influenced by the state of the labour market, and was highest where the demand for technicians was highest, namely in the South-East and North-West regions.

Membership of a trade union was mainly sought by ex-apprentices who had been promoted to the technical staff. Relatively few technicians belonged to professional organizations, largely because they did not possess the necessary qualifications for membership of such bodies. Most of those who aspired to join a professional association as an alternative to a trade union had been trained within the company concerned and wished to achieve a more general recognition of the skills they had acquired.

The most important factors leading to unionization were the ambiguity and uncertainty felt by many technicians about their career pattern and future place in the labour market. Most unionized technicians had been motivated by instrumental rather than ideological considerations. They were concerned with using bargaining strength to improve or restore their absolute and relative levels of earnings. In particular they wanted to see their own perceptions of their status compared with that of shop-floor workers rewarded equitably. At the same time they favoured salary payment schemes which were based upon rewarding individual merit.

Those who had joined trade unions had been influenced by their perceptions of powerful and militant manual workers' trade unions. Unionization was seen as a means by which technical staff could emulate the successes of the manual workers and achieve the restoration of the traditional salary differentials. This view of the possible achievements of technicians' unions was assiduously fostered by the unions seeking to organize technicians.

During the 1960s these issues were dramatized by a number of strikes in which white-collar workers were involved. Draughtsmen in particular showed a growing readiness to resort to strike action, but it is likely that threats of strike action by such eminently professional and middle-class groups as doctors had also had a significant influence on making strike action respectable to aspiring technicians.

Another factor giving rise to considerable unease and sometimes stimulating a sharp response was the threat or actual experience of redundancy. The rationalization and restructuring of industry brought about by mergers not only posed a threat to future job security, but also might compel mobility to a greater extent than with wage earners. It was, in particular, disruptive of the special relationship which technicians believed existed between themselves and management. Although technicians were generally satisfied with the technical competence of management they were deeply disappointed in the failure of management at the higher level to show sympathy for, and understanding of, their plight. Technicians were for the most part of the opinion that management was more ready to satisfy the demands of manual workers for higher pay and fringe benefits because they were organized in strong trade unions than they were disposed to look after the interests of technicians.

Although the basic factors stimulating trade-union development are to be found in the context of economic and social change, these factors do not explain why a particular establishment becomes unionized. In this respect other variables have to be taken into account, including the attitude of management to union recognition; the effectiveness of union organization; the legal obligation; and the cultural norms. More specifically, the existence of leaders amongst the dissatisfied group is an essential prerequisite of organization.

Some students of industrial relations have argued that successful union recruitment follows from management recognition and approval of unions. The findings of this study only partially substantiated this view. Although some relationship between membership of a trade union and a belief that management looked favourably upon the union was found, the most highly unionized group (laboratory staff) believed management to be opposed to their union allegiances. A high proportion of non-trade unionists maintained that they did not know whether management approved or not of staff members forming a trade union. The policy of the Government with regard to recognition is relevant in two respects. Firstly, as the major employer in the economy it is able to set an example to the rest of industry by giving union recognition to its white-collar staff employees in the hope that the private sector will follow suit. Secondly, the Government can introduce legislation that lays down a procedure whereby white-collar workers can obtain recognition for their union from the employer, on the pain of some sanction being imposed on an employer who refuses recognition.

Management has found itself having to adjust to a new situation, in which it is often no longer possible to sort out the problems of technical staff by a 'personal chat'. In the past, it had assumed that technical staff would identify with it and automatically understand and accept managerial decisions. The study showed that the managements of most of the companies involved were at least aware they had a problem with their technical staff, and most of them were taking some action to try and remedy the situation.

As a solution to the lack of career-advancement opportunities, management in some cases was adopting a policy of job enrichment and enlargement. It was hoped that by extending jobs by the addition of extra tasks and responsibility this would give the employees a greater sense of achievement and go some way to satisfying the aspiration of technicians for promotion and more control over their work. One company had the belief that the career advancement problem could be overcome by only taking on employees with a low level of aspiration. In an attempt to overcome salary dissatisfaction, managements were introducing job-evaluated salary structures for technicians. These managerial techniques, when successfully applied, to some degree provide a barrier to white-collar unionism since they cut down

the dissatisfaction manifested by technicians within the enterprise.

The response of technicians to the situation in which they
found themselves as revealed by this research may be summar-
ized as follows.

It is apparent that technicians are generally dissatisfied;
  (a) as individuals, in terms of their career opportunities, and
      their future opportunities for self-development and growth
      in personal responsibility and autonomy
  (b) as a group, with their present status and treatment as an
      occupational category within the firm. This discontent is
      largely expressed in market terms, most particularly in the
      unfavourable comparisons made between the peer or
      colleague group to which they belong and shop-floor earn-
      ings and conditions.

The very heterogeneity of the group which has created many
of the present problems also militates against universal solutions.
Clearly many of these problems are associated with the transition
from a situation in which the status of technician was significantly
associated with that of technologist or management, and one in
which to many technicians it was a mere step on the way to one
or other of these higher positions. The expansion of higher
education and the advance of modern technology has increased
the supply and demand for college-trained labour, which has had
the effect of throwing up higher and more formally defined
barriers to the upward mobility of technicians. For the present
generation of technicians whose college and workshop training
does not match up to the levels of education set by these new
hurdles, the next few years are bound to be a period of the utmost
frustration. Adjustment will not be easy, and there appear to be
few panaceas; the problems of adjustment are unique to each
group. The laboratory assistant with relatively high school-leaving
qualifications and a relatively undemanding job will react in a
different manner from the line technician with highly specific
skills, who sees himself to be in a relatively exposed and weak
position.

On the other hand there is little about the *ad hoc* way in which
the existing system of recruitment and induction is carried out
in most firms to commend itself as a model for the future. In
many organizations little attempt is made to treat technicians'
problems on an *occupational* basis, and in many instances the

'individual' treatment received by each person within the grade is an alternative to proper consideration of the recruitment and development of manpower at this level in a more integrated and long-term manner. At periods of organizational change it becomes most apparent that the attention that is paid to action in respect to shop-floor workers or to senior management is not given to the future of technicians as a group or occupational stratum. Since managers have not seemed to have regard to the status and security needs of these groups during periods of change and uncertainty, they seek their own forms of insurance in external unions.

It seems unlikely that the historic process that brought powerful craft unions into being 150 years ago can now be reversed in the technicians' case. While many firms had experienced an abatement in union activity after an initial protest action, the formalization of some grievance procedure for the technician group as an entity had become necessary in most large complex organizations. Some employers had demonstrated their ability to meet both the individual and group needs of technicians in a relatively successful manner, but the underlying market insecurity felt by many technicians seemed to suggest that the need for an external advocate would become more and more common in a world of redundancies and rationalization.

Yet for these representative bodies the next decade will be an extremely challenging one. Expectations built up over several generations of technicians will continue to be reflected in their early training and in the motives which lead them to select a technician career. Technicians will continue to demand recognition as a highly prestigious and largely autonomous group within the place of work, but the objective factors in their employment situation will continue to reduce their career opportunities and their relative status as a group. The education system, now gearing itself to the requirements of the technical college students, may eventually create a work force with more 'realistic' aspirations, but at the present time the evidence is that the more able among young potential technicians are attempting to opt out of vocational education and into more academic forms of socialization for their working career. If this is the case then the number of formally well-qualified youngsters entering technical careers may become so great that the newly created 'technician engineer' qualifications will lose their value almost as soon as they have come into existence.

Whether this is so or not, it seems probable that the basic social urge to differentiate one's membership group from that of others may lead technician engineers to seek professional status and the higher rewards of cash and power that accompany it. The movement to form professional-type bodies may therefore grow stronger as the pressures towards bureaucratic homogeneity grow more powerful. Yet ultimately it seems on the basis of historical evidence that the bureaucracy is likely to achieve control of the technology. In this event the movement towards collective bargaining seems inexorable. The course is not, however, by any means a smooth one, and the individual needs of technicians are and will continue in the foreseeable future to be closely related to their own career aspirations.

For the type of multi-group industrial or general union that is now emerging in the field of technician recruitment, these membership aspirations will continue to generate major substantive issues in terms of the service being offered to them, as well as being a source of weakness. The higher aspirations, as well as the objectively defined career opportunities for technicians, will continue to differentiate them from the manual-worker members of the A.U.E.W. and other organizations attempting to service both manual and non-manual workers. For the A.S.T.M.S. the problem is to be seen in terms of the increasing social distance growing up between white-collar workers themselves. By the end of the decade the majority of employed workers will be in white-collar occupations. Many lower-status clerical workers are already being treated on a par with semi- or even un-skilled manual workers. The cultural lag which enables these groups, as well as technician groups, to retain their self-perceived status may well work itself out over this period. For an organization that purports to represent professional and executive employees the decision to rest its bargaining strength on lower-status workers is a difficult one to take. There is at least one precedent for this, NALGO, but there is no staff union that has covered so many sectors of employment as the A.S.T.M.S. For this union the problem of meeting the quasi-professional aspirations of its technician and management groups while relying on lower-status members for its 'strength of numbers' will be a real test of organizational skills.

Bibliography

Figures in bold are the pages in this book on which the reference is made.

BOOKS

C. Argyris, *Personality and Organisation* (Harper, New York 1957), **141**

G. S. Bain, *The Growth of White Collar Unionism* (Oxford University Press 1970), **10, 11**, Chaps. 2 and 8, **52**, p. 34 Table 3.8, **59, 79, 87, 92, 272, 276**

J. A. Banks, *Industrial Participation: Theory and Practice, A Case Study* (Liverpool University Press 1963), **151**

G. S. Becker, *Human Capital* (National Bureau of Economic Research, New York, 1964), **128, 133**

P. Berger (ed.), The Human Shape of Work (Macmillan, New York 1964), **43**

R. M. Blackburn, *Union Character and Social Class*, (Batsford, London 1967), **52, 249**

M. Blaug (ed.), *The Economics of Education* I (Penguin Modern Economics Series, 1968), **133**

R. Blauner, *Alienation and Freedom* (University of Chicago Press, 1964), **154, 159**

M. Brown, *On the Theory and Measurement of Technical Change* (Cambridge University Press, 1966), **3**

T. Burns and G. M. Stalker, *The Management of Innovation* (Tavistock Publications, London 1961), **158**

Richard E. Caves et al., *Britain's Economic Prospects*, (Allen and Unwin, London 1968) pp. 448–484, **5, 23**

N. W. Chamberlain and J. W. Kuhn, *Collective Bargaining*, second edition, (McGraw-Hill, New York 1965), **59**

J. Child, *British Management Thought: A Critical Analysis* (Allen and Unwin, London 1969), **276**

330   RELUCTANT MILITANTS

S. Cotgrove, *Technical Education and Social Change* (Allen and Unwin, London 1958), **128**
Robert Dubin, *The World of Work* (Prentice Hall, Englewood Cliffs 1958), pp. 254–258, **138**
A. Etzioni, *A Comparative Analysis of Complex Organisations* (Free Press of Glencoe, New York 1967), **98, 139, 237**
G. Friedmann, *The Anatomy of Work* (Heinemann, London 1961), Chapters III and IV, **157**
J. Goldthorpe *et al.*, *The Affluent Worker: Industrial Attitudes and Behaviour*, (Cambridge University Press, London, 1970), **254, 264**
F. Herzberg, B. Mausner and Barbara Synderman, *The Motivation to Work* (Wiley, New York 1959), **50, 141, 254, 304**
J. Klein, *Samples from English Cultures* (Routledge and Kegan Paul, London 1965), **242**
Archie Kleingartner, *Professionalism and Salaried Worker Organization* (Industrial Relations Research Institute, The University of Wisconsin 1967), **250**
R. Layard, J. King and C. Moser, *The Impact of Robbins* (Penguin Educational Specials, London 1969), **27**
David Lockwood, *The Black-coated Worker* (Allen and Unwin, London 1958), **11, 52, 232**
K. Marx, *Das Kapital* (Foreign Languages Publishing House, Moscow), **251**
R. K. Merton, *Social Theory and Social Structure* (Free Press of Glencoe 1957), Chapters IV and V, **48, 115**
G. Millerson, *The Qualifying Associations* (Routledge and Kegan Paul, London 1964), **92, 249**
C. Wright Mills, *White Collar: The American Middle Classes* (New York: Oxford University Press, 1951), **237**
J. E. Mortimer, *A History of the Association of Engineering and Shipbuilding Draughtsmen* (A.E.S.D., 1960), **17, 77, 81, 85**
H. S. Parnes, *Research on Labour Mobility* (Social Science Research Council, New York 1954), **202**
K. Prandy, *Professional Employees: A Study of Scientists and Engineers* (Faber and Faber, London 1965), **52, 87, 88, 95, 268**
G. L. Reid and D. J. Robertson (eds.) *Fringe Benefits, Labour Costs and Social Security* (Allen and Unwin, London 1965), **194**
Lloyd G. Reynolds, *The Structure of Labor Markets* (Harper, New York 1951), **202**
Lloyd G. Reynolds, *Labor Economics and Labor Relations*, (Prentice-Hall, Englewood Cliffs, N.J., 1956), pp. 169–177, **64, 276**
John W. Riegel, *Collective Bargaining as Viewed by Unorganized Engineers and Scientists* (Bureau of Industrial Relations, Michigan University, Ann Arbor, 1959), p. 103, **46, 254**
Derek Robinson (ed.), *Local Labour Markets and Wage Structures*

(Gower Press, London 1970), **204**

G. Routh, *Occupation and Pay in Great Britain* 1906–60 (Cambridge University Press 1965), **10**

L. R. Sayles, *Behaviour of Industrial Work Groups*, (Wiley, New York 1958), p. 5 ff., **270**

Joseph A. Schumpeter, *Capitalism, Socialism and Democracy* (Allen and Unwin, London 1961), **5**

D. Seers, *The Levelling of Incomes since 1938* (Blackwell, Oxford 1951), **169**

E. L. Trist *et al.*, *Organizational Choice* (Tavistock Publications, London 1963), **157**

A. Touraine *et al. Workers' Attitudes to Technical Change* (O.E.C.D., Paris 1965) pp. 16–21, **159**

H. A. Turner, *Trade Union Growth, Structure and Policy*, (Allen and Unwin, London 1962), **57, 89**

P. F. R. Venables, *Technical Education* (Bell, London 1956) p. 239, **23**

Sidney and Beatrice Webb, *The History of Trade Unionism* (Longmans, London 1920), **51, 59, 284**

Sidney and Beatrice Webb, *Industrial Democracy* (Longmans, London 1919 edition) pp. 453–481 and pp. 704–714, **22, 36, 51, 57, 59, 80, 232**

M. Weber, *The Theory of Social and Economic Organization*. Translated by A. M. Henderson and T. Parsons. (New York: Oxford University Press 1947), **115**

N. de Witt, *Soviet Professional Manpower* (See an article in *The Times Educational Supplement* 30th December 1955; *Scientists and Technologists in U.S.S.R.*, by Prof. S. Zuckerman), **25**

Joan Woodward, *Industrial Organization: Theory and Practice* (Oxford University Press, London 1965), **123, 157**

ARTICLES IN BOOKS

R. Blauner, 'Work Satisfaction and Industrial Trends in Modern Society' in Walter Galenson and S. M. Lipset (eds.), *Labor and Trade Unionism*, (Wiley, New York 1960) p. 343, **138**

W. M. Evan, 'On the Margin: The Engineering Technician', in P. Berger (ed.) *The Human Shape of Work*, (Macmillan, New York, 1964), **38**

H. H. Gerth and C. W. Mills (eds.). *From Max Weber Essays in Sociology*, (Oxford University Press, New York, 1946) Chap. VII, **232**

E. C. Hughes, 'Licence and Mandate', *Men and their Work*, Chap. Six (Collier-Macmillan, Glencoe, London 1964), **58**

H. H. Hyman, 'The Value Systems of Different Classes', in R. Bendix and S. M. Lipset (eds,). *Class Status and Power* 2nd edition (Routledge and Kegan Paul, London 1967), **260**

R. K. Merton, 'Bureaucratic Structure and Personality', *Social Theory and Social Structure* (Free Press, Glencoe, Ill., 1957), pp. 195–206, **115**

Merton J. Peck, 'Science and Technology', in R. E. Caves *et al.*, *Britain's Economic Prospects* (Allen and Unwin, London 1968), Part X, pp. 448–484, **5**

J. H. Smith, 'The Analysis of Labour Mobility', in B. C. Roberts and J. H. Smith (eds.), *Manpower Policy and Employment Trends* (L.S.E.–Bell, London, 1966), **52, 147, 202**

Sidney and Beatrice Webb, 'The Method of Collective Bargaining', *Industrial Democracy* (Longmans, London 1919), Part II, Chap. II, **57**

R. K. Merton, Chaps. IV and V, *Social Theory and Social Structure* (Free Press of Glencoe 1957), **48**

ARTICLES IN JOURNALS

G. S. Bain, 'The Growth of White-Collar Unionism in Great Britain', *British Journal of Industrial Relations*, Vol. IV, No. 3 (November 1966), **10, 12**

R. M. Blackburn and K. Prandy, 'White-Collar Unionization: A Conceptual Framework', *British Journal of Sociology*, Vol. XVI, No. 2 (June 1965), **232, 234**

L. Broom and J. H. Smith, 'Bridging Occupations', *British Journal of Sociology*, December 1963, **203**

J. R. Bright, 'Does Automation raise skill requirements?' *Harvard Business Review*, Vol. 36, No. 4 (1958) pp. 85–98, **19, 159**

Burton R. Clark, 'The "Cooling-Out" Function in Higher Education' *American Journal of Sociology*, (May 1960), pp. 569–576, **23**

Ronald G. Corwin, 'The Professional Employee: A Study of Conflict in Nursing Roles', *The American Journal of Sociology* (May 1961), pp. 604–15, **54**

S. Cotgrove, 'Technicians in Industry', *Technology*, November and December 1960, **42**

S. Cotgrove, 'Does the Boss Care', *Technology*, August 1962, **42**

S. Cotgrove, 'The Relations between Work and Non-Work among Technicians', *Sociological Review*, Vol. 13, No. 2 (New Series) July 1965, **42, 46, 242**

E. Domar, 'On the Measurement of Technological Change', *Economic Journal*, December 1961, p. 1, **3**

J. C. R. Dow and L. A. Dicks-Mireaux, 'The Excess Demand for Labour', *Oxford Economic Papers*, February 1958, **208**

M. Fogarty, 'The White-Collar Wage Structure', *Economic Journal*, 1957, **172**

J. F. B. Goodman, 'The Definition and Analysis of Local Labour Markets: Some Empirical Problems', *British Journal of Industrial Relations*, Vol. VIII, No. 2, (July 1970), **204**

A. W. Gouldner, 'Cosmopolitans and Locals: Towards an Analysis of latent social roles', *Administrative Science Quarterly*, Vol. II, Nos. 1 and 2, (1957–1958), **206**

G. K. Ingham, 'Organizational size, orientation to work and industrial behaviour', *Sociology*, 1 (1967), pp. 239–258, **155**

Archie Kleingartner, 'The Organisation of White Collar Workers', *British Journal of Industrial Relations*, Vol. VII, No. 1, (March 1968), **91**

K. G. J. C. Knowles and D. J. Robertson, 'Differences in Wages between Skilled and Unskilled Workers, 1880–1950' *Bulletin of the Oxford University Institute of Economics and Statistics*, Vol. 13, No. 2 (April 1951), **9**

Kurt Lewin, 'The Psychological Effect of Success and Failure', *Occupations*, XIV (1936), pp. 926–930, **156**

K. Macdonald and W. A. T. Nichols, 'Employee Involvement: A Study of Drawing Offices', *Sociology*, Vol. 3, No. 2, (May 1969), **46**

A. H. Maslow, 'A Theory of Human Motivation', *Psychological Review*, Vol. 50, 1943, pp. 370–396, **50**

Louis H. Orzack, 'Work as a "Central Life Interest" of Professionals', *Social Problems*, Vol. 7, Fall 1959, **46, 254**

William J. Paul, Keith B. Robertson and F. Herzberg, 'Job Enrichment Pays Off', *Harvard Business Review*, March/April 1969, **304**

D. S. Pugh, 'Role activation conflict: a Study of Industrial Inspection', *American Sociological Review*, Vol. 31, No. 6, (December 1966), **153**

D. Pym, 'Education and Employment Opportunities for Engineers', *British Journal of Industrial Relations*, Vol. VII No. 1 March 1969, **302**

R. L. Raimon, 'Changes in Productivity and Skill-Mix', *International Labour Review*, Vol. 92, No. 4, (October 1965), **9**

W. B. Reddaway, 'Wage Flexibility and the Distribution of Labour', *Lloyd's Bank Review*, (December 1959), pp. 32–48, **202**

D. Silverman, 'Clerical ideologies – A research note', *Sociology*, June 1968, pp. 327–333, **165, 266**

A. L. Stinchcombe, 'Bureaucratic and Craft Administration of Production', *Administrative Science Quarterly*, Vol. 4, 1959–60, pp. 168–187, **22**

Graham Wootton, 'Parties in Union Government: the A.E.S.D.', *Political Studies*, Vol. 9, June 1961, **83**

Robert Oubin, 'Industrial Workers' Worlds. A Study of the "Central Life Interests" of Industrial Workers', *Social Problems*, Vol. 3, (January 1956), p. 131, **3, 38**

D. E. Weir and O. T. H. Mercer, *Report on Social Science Research Council Conference on Industrial Relations and Social Stratification*, mimeo (Cambridge 1969), **151**

GOVERNMENT PUBLICATIONS

Committee on Manpower Resources for Science and Technology, *Report on the 1965 Triennial Manpower Survey of Engineers, Technologists and Scientists and Technical Supporting Staff*, Cmnd. 3103 (H.M.S.O., London, October 1966), **18, 21, 42, 128**

Employment and Productivity Gazette, January 1970, **10**

Statistics on Incomes, Prices, Employment and Production (ceased June 1969), **207**

Ministry of Labour, 'Annual Survey of Occupations in Manufacturing' May 1964, May 1968, *Ministry of Labour Gazette*, January 1965 January 1969, **15**

Ministry of Labour, 'Survey of the Employment of Technicians in the Chemical and Engineering Industries', *Ministry of Labour Gazette*, December 1960, **13, 18**

Ministry of Labour Gazette, January 1966, p. 20, **123**

Percy Committee Report, *Higher Technological Education* (H.M.S.O., London 1945), p. 6, **25**

Report of the Committee on Technician Courses and Examinations, (Dr. H. L. Haslegrave), National Advisory Council on Education for Industry and Commerce, Department of Education and Science (H.M.S.O., London 1969), **22, 29, 30**, Table 2, pp. 13–14, **27** Paras. 224–226, **32**, Para. 270, **34**, Para. 8 p. 5, **39**, Para. 20 p. 7, **40**, Para. 22. p. 8, **41, 42**

Report of a Court Inquiry under Lord Pearson into the dispute between the B.S.C. and certain of their employees, Cmd. 3754 (H.M.S.O., London, August 1968), **169, 109**

Report of the Chief Registrar of Friendly Societies for the year 1969, part 4 'Trade Unions' (H.M.S.O., London 1970), **60**

Report of a Committee appointed by the Council for Scientific and Industrial Research (Chairman, G.B.R. Feilden), *Engineering Design*, (H.M.S.O., London 1963), **8, 21, 34**

Technical Education, Cmd. 9703 (H.M.S.O., London 1956) p. 2, **42**

'The Survey of Professional Engineers 1968', *The Survey of Professional Scientists, 1968*. Ministry of Technology (H.M.S.O., London 1970), **74**

Advisory Council on Scientific Policy. Committee on Scientific Manpower, *The Long-Term Demand for Scientific Manpower*, Cmnd. 1490 (H.M.S.O., London 1961), **7**

Memorandum on Higher Technological Education (H.M.S.O., London 1956), **26**

National Advisory Council on Education for Industry and Commerce, *The Future Development of Higher Technological Education*, Ministry of Education (H.M.S.O., London 1969), **25**

Ministry of Labour, *Occupational Changes 1951–61*, Manpower Studies No. 6 (H.M.S.O., London 1967), **11**

Ministry of Labour, Manpower Studies No. 2, *The Metal Industries*, (H.M.S.O., London 1965), **13, 16, 20**

Department of Employment and Productivity, *Report on Technician Training*, (H.M.S.O., London 1970), **34**

National Board for Prices and Incomes, Report No. 49, *Pay and Conditions of Service of Engineering Workers* (1st Report), Cmnd. 3495 (H.M.S.O., London, December 1967), **168, 189, 196, 199, 308** and the *Statistical Supplement*, Cmnd. 3495-1 (H.M.S.O., London 1968), **168, 196, 199**

National Board for Prices and Incomes, Report No. 68, *Agreement made between certain Engineering Firms and DATA*, Cmnd. 3632 (H.M.S.O., London, May 1968) paras. 34–40 and Appendix E, **168, 192, 309**

National Board for Prices and Incomes, Report No. 86, *Pay of Staff Workers in the Gas Industry*, Cmnd. 3795 (H.M.S.O., London 1968), **309**

The Report of the Committee on Scientific Manpower (Barlow) 1946, p. 10, **6**

Better Opportunities in Technical Education p. 5, Cmnd. 1254 (H.M.S.O., London 1961), **26, 42**

A Report of Central Advisory Council for Education (England), under Chairmanship of Sir Geoffrey Crowther, Ministry of Education, 2 vols. (H.M.S.O., London 1959), **23**

Half our Future, Central Advisory Council for Education, Department of Education and Science (H.M.S.O., London 1963), **23**

ROYAL COMMISSION ON TRADE UNIONS AND EMPLOYER ASSOCIATIONS

G. S. Bain, *Trade Union Growth and Recognition*, Research Paper No. 6, Royal Commission on Trade Unions and Employers' Associations (H.M.S.O., London 1967), **79, 87, 268**, pp. 68–97, **315**

Alan Fox, *Industrial Sociology and Industrial Relations*, Research Paper No. 3, Royal Commission on Trade Unions and Employers' Associations (H.M.S.O., London 1967), **316**

Royal Commission on Trade Unions and Employers' Associations, Minutes of Evidence No. 36, D.A.T.A. (H.M.S.O., London 1966), **75, 82, 83, 192, 232**, Appendix 59–63, **314**

Royal Commission on Trade Unions and Employers' Associations, 1965–1968. Report, Cmnd. 3623, (H.M.S.O., London 1968), Para. 252, **65**

Royal Commission on Trade Unions and Employers' Associations, Minutes of Evidence No. 53 ASSET (H.M.S.O., London 1967), **97, 99, 113**

EMPLOYERS

Report on the Education and Training of Technicians in Laboratory and

Quality Control Departments, (British Iron and Steel Federation, October 1962), **43**
Report on the Recruitment and Training of Technicians, (British Iron and Steel Federation, April 1961), **43**

TRADE UNION SOURCES
A.S.T.M.S. Journal, Issue 3, 1969, **117**
British Employers' Confederation, 'Wages and Conditions–Exchange of Views and Information. Staff Workers, June 1964', DATA's Evidence Appendix 59–63, **314**, p. 61–2, **314**, p. 62–3, **314**
DATA Journal, June 1968, **110**, August 1968, **110**, November 1967– Précis of article **110**, May 1970 'U.K.A.P.E. – unpublished reply' p. 19, **113**
A Declaration of Dissent: The Technicians Union — Incomes Policy (1965), Chap. 6, **189**, *The Draughtsman*, October 1968, **105**, January 1968, **107**, April 1968, **107**, June 1968 Annual Conference Report, **107**
'Managers, Supervisors, Technicians: Your Future and status' (ASSET, 1946), **101**

NEWSPAPERS
Daily Telegraph, June 1967, **103**
Financial Times, 12th June 1969, **71**, 16th January 1970, **72**
The Guardian, 15th April 1968, **114**, 9th May 1968, **110**, 2nd May 1969, **110**
S. G. Cotgrove, *New Scientist*, 20th November 1958, pp. 1308–1310, **300**
New Society, 30th May 1963, **103**, **105**
The Times, 14th July 1967, **71**, 7th November 1969, **73**, 28th September 1970, **72**, 27th October 1970, **72**, 23rd June, 1970, **72**, 8th July 1970, **73**
The Times Educational Supplement, 30th December 1955, **25**
Sunday Times, 16th August 1970, **71**, 1st December 1968, **102**
'Technology: The Revolution that Wasn't', *The Economist*, 19th March 1966, p. 1147, **5**

OTHER
G. S. Bain, (quoted) from a report of an observer to the Engineering Employers' Federation, **99**
J. V. Eason and R. Loveridge, 'Management Policy and Unionisation among Technicians', in *Proceedings of a Social Science Research Council Conference on Social Stratification and Industrial Relations* (Social Science Research Council, Cambridge 1969), **155, 256, 269, 284**
'Investigation concerning Technicians associated with automated processes', *U.K.A.C. Record*, No. 5, February 1965, **42**

C. Jenkins, 'Tiger in a White Collar'. *Penguin Survey of Business and Industry 1965*, **103, 104**

Joint Committee on Practical Training in the Electrical Engineering Industry set up by the Institution of Electrical Technician Engineers, **45**

Manpower (Political and Economic Planning, London 1951), **17**

Recommendations on the Training of Technicians and Technologists, (Iron and Steel Industry Training Board, May 1966), **43, 43**

The Education, Training and Functions of Technicians: United Kingdom (Directorate for Scientific Affairs, O.E.C.D., Paris 1966), **41**

Training of Chemical, Metallurgical and Fuel Laboratory Assistants and other equivalent jobs (Iron and Steel Industry Training Board 1968), **43, 44**

Report of an enquiry into the employment and training of technicians in the Engineering Industry, **40**

Professional Engineering Responsibility Levels (Professional Engineering Publications, Liverpool), **74**

Recommended Salary Levels for Professional Engineers (Professional Engineering Publications, Liverpool), **74**

Alexander Wedderburn, '*Work Orientation among Draughtsmen*' Social Science Research Council Conference in Social Stratification and Industrial Relations, Cambridge (unpublished), **46**

U.K.A.C. Record, No. 9, p. 9, **41**

R. A. Ulrich, 'A theoretical model of human behaviour in organizations: an eclectic approach', unpublished dissertation (London School of Economics 1969), **148**

Index

shop-floor differentials, attitudes
   215, 216, 217, 262, 264, 307,
   320–2
status 46, 53
   marginal 320
   and salary 58
   views on 132, 145, 147
substitution for technologists 134
survey of 121–4
   results 124–7
treatment by management 276
unionization 56, 324, 323–7
work-place attitudes 132

Technician engineers 33, 54, 70
   loss of value of 327–8
Technicians' institutions 56
Technological changes 3
   and labour demand pattern 4
   and work environment 165
Technological degrees 25
   effects of development of 27
Technologists 4
   shift to college education 24
   substitution of 8
   technicians employed as 205
Technology, effect on behaviour of
   technicians 156
Trade unions, attitudes to 231
   "class" explanation 251
   inappropriate to technicians 315–6
   open 57
      technicians in 59, 284–7

recognition by management 269,
   271, 316–7
   types of 313
Trade Union members, and professional associations 240–1
   unambitious 257
Training in industry 32
   wastage rate 23
Transport and General Workers'
   Union, technicians in 63

Unemployment figures, as source of
   statistics 207
Unionization 234
   and education 244–5
   and family background 242–3
   process of 283
   and qualifications 251, 253
   and salary levels 188
   and recruitment and training 246
   reduction in growth of 272
   among technicians 56, 186–7, 234,
      323–7
United Kingdom Association of Professional Engineers (UKAPE)
   70, 71
   as trade union 71

Woodward, Joan 157–8, 160
Work group, effects on unionization
   277–8
Work group relationships 256
Work-study engineers 154